BACKPACKING
in the 80's

BACKPACKING
in the 80's

by

Bob Wirth

Parker Publishing Company, Inc.

West Nyack, New York

© 1984, by

PARKER PUBLISHING COMPANY, INC.
West Nyack, N.Y.

Library of Congress Cataloging in Publication Data

Wirth, Bob,
  Backpacking in the 80's.

  Includes index.
  1. Backpacking.   2. Backpacking—Equipment and
supplies.   3. Backpacking—United States.   I. Title.
II. Title: Backpacking in the eighties.
GV199.6.W57   1984   688.7'65'1   83-13249
ISBN 0-13-056747-7
ISBN 0-13-056739-6 {PBK}

Printed in the United States of America

For T.J., whose anonymous note left at the top of a mountain reflects the spirit of the great outdoors. . . .

.. Scorpio's tail shows to the South. Cygnus' wing touches the eastern horizon. And the moon, short of First Quarter by a day, is preparing its descent to the West. And here am I, alone. Once again on this volcanic prow that thrusts itself into the desert sky. And I'll keep coming back, too. You'll whittle me down bit by bit, tooth by tooth. But I'll be back. In the end, I'm all yours anyhow. I wouldn't want it any other way.

The moon is down. I can see the tiny points of light where the city rests so quietly, so distant. Another year, old Mountain — another year. You've been good to me. Rest well. I'll be back . . . . .

T. J.

How This Book
Can Help You

Backpacking. Just the word itself creates images of carrying a heavy pack up a steep mountain, of freezing in a cold sleeping bag, of walking miles in a torrential downpour, and of rattlesnakes, scorpions, bears, and mosquitoes. Misery in all its glory. Almost. That word also creates images of the sky painted with a thousand morning colors, of 100-mile vistas, of friends sharing a campfire, and of the proud feeling that comes when you know you are where you are because you walked there.

Backpacking is all those things and it's booming. It seems like its popularity is doubling every year. In a few short years, backpacking changed from a sport that involved only a handful of Boy Scouts to a sport with millions of people participating. New outdoor sports stores and mail-order suppliers are opening up across the country to meet the demand for camping gear and they are marketing a host of bewildering new equipment designed specifically for people interested in backpacking. Materials such as polypropylene and pile and equipment like lightweight hiking boots have revolutionized the sport in the last few years. Backpacking in this decade is indeed far different than it was in the 1970's.

If you are new to backpacking, everything in this book will help you. Read it, re-read it, go hiking, and then read it again. Although it's written in a clear, easy-to-follow format, the concepts presented in it will make a lot more sense to you after you have hiked a few times. If you have prior backpacking experience, this book may surprise you. Backpacking is far different than it was even a few years ago. Heavy leather boots are obsolete. Pile is far better than wool in wet conditions. Vapor barriers do work in cold weather. The popular cut-and-suck method of first aid for snakebites is often more dangerous than the venom itself. Indeed, *Backpacking in the 80's* is for people just getting interested in the sport as well as for experienced hikers who think they know what they are doing.

Backpacking in the 80's is a complete "how-to" book describing all aspects of backpacking in an easy-to-read, tight, non-preachy, person-to-person style. As you read it you'll get the feeling that I'm talking directly with you. You'll feel that I'm explaining very technical terms in a nontechnical, easy-to-understand fashion. However, don't be misled by its comfortable writing style. There are no filler words or off-the-track stories in this book. Though it is clear in reading, it is concise in content. A few words say a great deal in the book.

This book is organized in a clear, simple, easily referenced fashion. If you are new to backpacking, simply read it from cover to cover and then re-read sections of it as needed. The most important items you need to know are introduced first. For example, you'll read about the kinds of equipment available before you'll learn about actually hiking with that equipment. Specialized outdoor skills that take a while to master, like knot-tying and using a compass, are explained later in the book. If you're an experienced backpacker, you can read the entire book from cover to cover at your leisure or sections of it as needed. For example, if you're in the market for a new sleeping bag, simply read that part of the equipment section. No matter how much hiking experience you have, though, you'll find the easily referenced table of contents and complete index useful for cross-referencing topics discussed in various sections of the book.

Unlike many other backpacking books, *Backpacking in the 80's* doesn't discuss specific products, brand names, or prices, since they change a great deal from year to year. Instead, it contains only the principles of how to buy equipment and lets you decide what models to buy based on your own needs. In addition, the book covers a range of techniques from the basics of walking to the complexities of planning a solo, long-distance hike and teaches important campcraft skills that range from building a fire to predicting the weather. *Backpacking in the 80's* is not a survival book but it will teach you how to survive outdoors. While it's written in clear, simple language and assumes you have no previous backpacking or camping experience, it's not written just for beginners. Because it contains information about backpacking equipment, hiking techniques, and campcraft skills needed for a variety of backpacking trips under a host of conditions, it's probably the only outdoor camping/hiking book you'll ever need. Happy trails!

Bob Wirth

CONTENTS

II

TECHNIQUES

III

CAMPCRAFT SKILLS

IV

DANGER

I

EQUIPMENT

1

Three Important Concepts

Because so much of backpacking equipment design is based on the concepts of weight, bulk, and insulation, let's look at them first.

Weight

Simply stated, the more weight you carry, the more time and energy you need to carry it. Thus, the lighter your pack is, the easier, faster, and farther you can hike in a given amount of time. The more your equipment weighs, the less distance you can hike and the more tired you will be at the end of each day. Weight is so critical that you need to be constantly aware of it when buying equipment, planning your hikes, and packing your gear.

Here are three ways to reduce the weight of your pack:

1) *Carry only lightweight items.* As you will soon find out, there are incredible differences in the kinds of equipment available for your use. For example, one type of sleeping bag weighs half as much as another yet keeps you just as warm. One backpack weighs several pounds less than

other models designed for identical purposes. Plastic water bottles weigh less than half as much as similar aluminum ones. In general, though, you pay for lighter gear with their greater expense and lower durability. This is especially important to understand in this age of ultra-light gear appearing on the markets.

2) *Reduce the number of things you carry.* Be a butterfly, not a packhorse. Simplify. Carry gear that has a variety of uses, and carry only what you truly need. For instance, leave your fork, dinner knife, and plate at home. Eat and drink from your cup with your spoon. "When in doubt, leave it out" experienced backpackers say. The fewer items you carry, the easier it will be to find the things you need in your pack, and the less you'll have that could be stolen, damaged, or lost on the trail. Leave all camera equipment, books, and other unneeded items at home. Don't get caught up with gimmicks and outdoor toys associated with backpacking. There is value in watching clouds drift across the sky for hours. Keep it simple, travel light, and stay free.

The only exception to this is when you are going on an easy, undemanding hike. Then, feel free to carry anything you want to, including a case of beer, a watermelon, and a huge steak. Having a good time outdoors is what backpacking is all about, whether you backpack to a remote wilderness fishing spot or to an overnight party on a neighbor's farm.

3) *Share as much gear as possible* with your hiking companions. For example, one flashlight will serve two to four people with only minor inconvenience. One small backpacking stove is all a group of five to eight people need on an extended outing if they don't mind cooking in two shifts. One large expedition tent full of people is much less weight per person than if everyone carried their own one- or two-person shelter.

Bulk

The things you carry should be small and compact as well as light. A small-sized, more compact backpack is easier to carry because its cargo cannot shift around as much on your back as you walk. It's easier to put on, easier to take off, easier to load and unload, and easier to carry around and over obstacles. Also, items you need are much easier to locate inside it. Walking cross-country with a tight, compact pack is much less tiring than hiking with an oversized one of equal weight. Bulk becomes extremely critical on long-distance hikes and in cold weather when you have to carry large amounts of supplies. Reduce the volume of your pack by reducing the number and size of the things you carry in it.

Insulation

Clothing and certain kinds of camping equipment like sleeping bags are the only protection you have from a frequently hostile and often unpredictable environment. While you can protect yourself from excessive heat by removing clothing, it's much more difficult to keep warm when the weather is cooler. In cool weather, *your body loses heat* in five ways:

1) *Conduction* is the direct transfer of heat from one object to another. You lose heat through conduction when you sit on a cold rock, for example.

2) Your body loses heat by *convection* when a breeze blows across it.

3) Your body loses heat through *evaporation* when water is transformed into water vapor. When your clothing is wet, the combined effects of that water conducting your body heat away and the water itself evaporating can be deadly.

4) Your body loses heat when your lungs warm cold air through *respiration*.

5) *Radiation* is heat lost to colder objects by infrared rays. For example, the warmth you feel from a campfire is radiated heat. In cold weather as much as three-quarters of your body's heat can radiate from your uncovered head.

All camping equipment and clothing designed to protect you from the elements is affected by those principles of heat loss. To stay warm, follow these *principles of insulation:*

1) Clothing doesn't keep you warm. It simply prevents your body warmth from escaping to a colder environment. *The best insulating substance is dead air space,* which is tiny pockets of air that cannot move. The material that traps the most dead air is the warmest, but the thickness of that material itself is not sufficient to guarantee warmth. A thick garment is not necessarily a warm garment, because research has shown that insulation depends more on a material's amount of surface area than on its thickness. Even a few millimeters distance between the insulating fibers within a garment is enough to form convection currents between them and reduce their heat-trapping ability (Example 1-1). Yet, while insulating fibers must be close enough to prevent convection heat loss, they should be far enough apart to prevent heat lost through conduction.

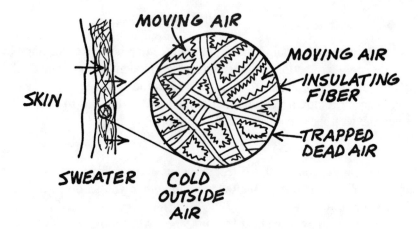

Example 1-1: Heat lost by convection through insulation.

Air is trapped and held in place by the friction generated when it touches the insulating fibers. Thus, the greater the surface area of the insulating material, the warmer it will be. That's why new materials like Thinsulate provide the same warmth value as twice as much down (by thickness), and clothes with a fuzzy weave like sweaters insulate better than clothing with a slick, tight weave like denim (Example 1-2). In other words, a light, loosely woven sweater could provide as much insulation as a heavy, tightly woven one if their fibers have the same surface area and trap the same amount of dead air.

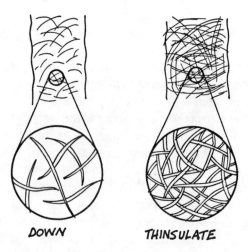

Example 1-2: A comparison of the surface area of down and Thinsulate insulation.

2) *Get out of the wind*. Prevent convection currents from carrying air trapped in your insulation away by wearing a windproof outer layer and sleeping in a windproof shelter like a tent. A slight breeze is enough to reduce a sweater's insulation value to almost nothing, for example.

3) *Stay dry*. Water conducts heat away from your body 240 times as fast as air does. Even water vapor trapped in the spaces between insulation fibers reduces the effectiveness of those materials.

4) *Ventilate* your clothing at neck, wrist, waist, chest, and ankle openings and remove several clothing layers before exercising so your sweat can evaporate instead of soaking your clothes. Because you breathe out over 1 pint of moisture each night, always keep your nose and mouth out of your sleeping bag when sleeping. If you breathe inside the bag, that moisture will collect in the insulation and chill you.

5) *Dress in many thin layers* instead of one thick layer. This traps additional air between the layers and lets you better adjust for changes in your metabolism. Simply remove a layer or two before exercising for better ventilation and add a layer or two after exercising to prevent chilling.

The layer of clothing closest to your skin should be very *breathable*, which means it allows moisture to pass through easily so it doesn't collect in your clothing and conduct heat away. Synthetic materials like polypropylene and clothing with a large-weave, fishnet design are ideal for this purpose. Depending on the temperature, the next several layers of insulation should trap as much dead air as possible near your body for warmth. These layers could include sweaters, vests, coats, or your sleeping bag. The final, outer layer should be windproof and water repellent to prevent wind from blowing the dead air out of your insulation and water from conducting your heat away. Examples include a parka over a sweater and a tent around your sleeping bag.

6) *Use products that reflect radiated heat rays* back to your body. For example, a rescue blanket (also called a "space" or "thermal" blanket) is just a specialized piece of aluminum foil that reflects your radiated heat back to you. A thin piece of reflective material alone won't keep you warm in cold weather, though, because your body will still lose heat through conduction, convection, evaporation, and respiration, as well as radiation.

7) *Material colors*. Dark-colored materials absorb heat, while light-colored ones reflect heat rays in the direction from which they came. When hiking in the desert in summer, wear light-colored clothing that reflects the sun's rays away from your body to keep cool, but when hiking in cold weather wear dark-colored clothing that absorbs as much of the sun's warmth as possible.

Condensation

Meteorologists tell us that the warmer air is the more water vapor it can hold, while the colder the air temperature is the less water vapor it can hold. During the day, as the sun warms the atmosphere, it absorbs evaporated water from the environment. Then, when that warm air cools in the evening, the air loses its capacity to hold the water. If it reaches the *dew point,* the temperature where the air is saturated with water vapor and cannot hold any more, excess water vapor turns to visible drops of water and condenses on the ground as dew. The greater the difference between the warm and cold temperatures and the greater the initial concentration of water vapor in the warm air, the greater the chance of the dew point being reached.

The same thing happens to your clothing and camping gear. Air near your body is warmer and usually well saturated with your evaporated body moisture. As that air moves from your warm body through your insulation layers to the cold outside air, it cools and can hold less moisture than it did near your body. If the air temperature reaches the dew point in or between your layers of insulation, the water vapor will condense into liquid water and soak your insulation in a very short time (Example 1-3).

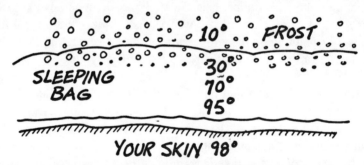

Example 1-3: Water vapor condensing to liquid water inside your insulation turns to frost when the temperature is below freezing.

Ventilating your insulation to prevent a buildup of water vapor there, controlling your physical activity to reduce the amount of water vapor given off as sweat, removing clothing layers before exercising, and wearing highly breathable clothing materials to let water vapor easily pass through them are the best ways to prevent this condensation in your insulation.

2

Obtaining Equipment

Before buying any camping equipment, *consider* the factors which determine the kind and amount of gear you'll need:

1) *What kind of trips you'll take.* Are you looking for inexpensive equipment for hikes a few miles from home or do you need top quality gear to climb rugged wilderness peaks? Do you need very lightweight equipment for long distance hiking, or are you interested primarily in short hikes where weight is a less critical factor?
2) *The predominant weather conditions.* Will you do most of your hiking in warm, cold, wet or dry climates, or do you need equipment for combinations of those conditions? If you hike mostly in the arid deserts, you might not need any shelter at all, but if you hike in the always wet Cascades, you'll need the best tent or tarp you can buy. You'll need a down jacket in the Rocky Mountains in summer, while a sweater will keep you warm on the coldest nights in southern Florida.

Next, *get advice* from as many different sources as possible. Read books, write for camping catalogs (see Appendix for addresses), talk to salesmen in stores, and join an outdoor club. If you don't know anyone who backpacks, drive to a state or national park and talk to the rangers or hikers there. Don't buy anything until you've thoroughly researched what's available, considered what you need, and know the going prices

23

for that equipment. The more you know about what kind of equipment you need and how much it sells for, the easier you'll recognize a bargain when you see one.

Then *make a list* of the gear you'll need according to your priorities. If you plan to camp mostly in the desert and mostly on short hikes, you'll need a day pack long before you'll need a tent. If you'll camp on the Long Trail ("the longest river in Vermont," the locals say), you'll need good boots before you'll need an extra canteen. Buy accordingly.

When beginning, *use as many things from home as possible.* Improvise. Use sneakers instead of boots, plastic bags for raincoats, and a gym bag for a backpack. When you do need to buy equipment though, follow these guidelines:

1) *Buy gear adequate for most of your trips most of the time.* For example, if you'll camp mostly in summer, it's better to buy a medium-weight sleeping bag for general use in moderate temperatures and carry an extra set of long underwear for camping on cold nights. Don't buy a very warm bag for the coldest temperatures you'll ever experience and sweat in it all summer.

2) *Buy the best equipment available for your needs* despite its cost, but remember that the most expensive gear is not necessarily the best for you.

3) *Buy what you need when you need it.* Don't stockpile expensive, little-used gear.

4) *Avoid ignorant sales clerks* in fashionable outdoor sports stores who are there to sell you merchandise and not to help you buy it. Avoid catalog suppliers with a poor reputation and return/guarantee policy. Shop around. Often thrift, discount, and army surplus stores sell a host of drab, unfashionable gear at down-to-earth prices.

5) Consider *renting equipment* if you can't afford it, if you plan to backpack only a few times a year, or if you want to test a particular item before buying it. You can rent almost any kind of gear including tents, packs, stoves, and sleeping bags, and many stores will credit your rental fee towards the purchase price if you decide to buy it. Rent from camping supply stores, college recreation departments, and army bases.

6) *Buy used camping equipment.* Check newspaper ads, college bulletin boards, and outdoor club newsletters for bargains. Advertise your desire to buy a specific kind of used gear by posting "wanted" signs in foodstores, newspapers, laundromats and on college bulletin boards. When buying used equipment, check it carefully for frayed seams, peeling waterproof coatings, patched fabric, and bent parts, all of which are

signs of heavy use, poor quality materials, or improper care. Figure out how much the gear is worth to you in good used condition, and then lower your price for any defects like broken zippers or holes you find in it.

7) If you have a sewing machine and some free time, you can save half the price of camping equipment by making it from a *kit*. Most kits have easy-to-follow directions that let you alter the design or add features to fit your needs. For example, you can add handwarmer pockets to clothing and sew a shoulder collar inside a sleeping bag. See the Appendix for a list of suppliers.

Children's Equipment

Obtaining equipment for growing children poses special, but not overwhelming, problems. Use as much equipment from home as possible, and buy only what you can't possibly get at home or from neighbors. Plan for growth when buying any children's equipment. Buy larger sleeping bags than necessary, adjustable pack frames, and inexpensive sneakers instead of easily outgrown hiking boots.

3

Backpacks

Kinds of Packs

A pack is a specialized bag or compartment used to carry your supplies. There are two main kinds.

A *day pack* is a small pack with no frame that's used for short hikes, or as a school book bag, a gym bag, or a backpack for small children on overnight trips. They can comfortably carry up to about 20 pounds of gear. Features of day packs include foam padding to keep sharp objects away from your back, padded shoulder straps for carrying comfort, zippered—not snapped—pockets for securely holding small objects, a strap (similar to a backpack hipbelt) securing the bottom of the pack across your chest when running with it on, and a drawstring closure to secure the main opening. All day packs have some of these features; few have them all. Buy according to your needs. A *fanny pack* is a specialized day pack that wraps around your waist. It can carry a few pounds of useful items like a camera, lunch, and a sweater. If your day pack is not padded, place soft items like clothing against your back for padding so sharp objects like a camera won't poke you in the back (Example 3-1).

Small *framed backpacks* can hold enough gear for an overnight camping trip, while larger ones are designed for extensive expeditions. They come in two main styles (Example 3-2).

Internal Frame Backpacks

Internal frame packs have frames inside of them. Because they ride closer to your back, have a lower center of gravity, are flexible, and allow

26

Example 3-1: Types of day packs.

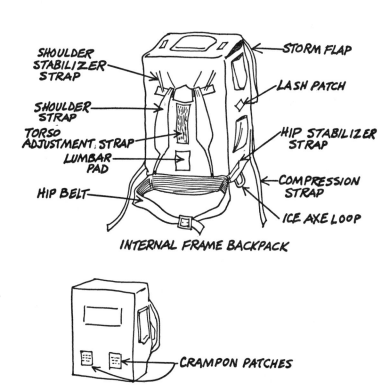

Example 3-2: Features of an internal and an external frame backpack.

EXTERNAL FRAME BACKPACK

Example 3-2: Features of an internal and an external frame backpack (Cont'd.).

more freedom of movement than external framed packs, they're well suited for cross-country skiing, snowshoeing, rock climbing, and off-trail exploring, as well as for general backpacking on trails. Many international travelers use them because they have few rigid parts and loose edges which snag in luggage terminals, while canoeists like them because they easily fit inside their boats. Although some models hug your back which allows sweat to collect and cause an itchy, hot, annoying feeling, newer and more expensive models overcame this problem by designing an air space between the pack and your back for better ventilation (Example 3-3). Internal frame packs generally fit better and feel more comfortable than external frame packs. Although they don't appear as tough as packs with external frames, quality internal frame packs are very durable and often have a lifetime guarantee.

Internal frame packs have reinforced fiberglass, plastic, or metal supports sewn inside pockets of fabric in an "11" or an "X" pattern (Example 3-4). Parallel supports are more flexible on your back, while X-shaped ones provide slightly more load support. In general an X pattern keeps a pack a little farther from your back so that it rides more like a separate unit on it (like an external frame pack does). While satisfactory for rugged, off-trail use, it is not specifically designed for that. Because an internal frame pack with parallel supports hugs your back closely and rides on it as if it were a part of it more than an X-framed one does, it is the best type of pack to use for cross-country skiing and off-trail ram-

Example 3-3: Internal frame pack with an air space for ventilation.

bling. This discussion is rather academic in nature, however. You may or may not notice these differences while hiking.

Because of their smaller carrying capacity and because the sleeping bag is usually carried inside these packs, they're ideal for hikes two to eight days in length. Despite manufacturing claims, only a select and expensive few internal frame packs are suitable for long-distance, expedition use. Many large-volume internal frame packs are over-designed for marketing purposes and are not built strong enough to carry heavy and large loads. Before buying a large internal frame pack, carefully check the places where the shoulder straps and hip belt are attached to the packbag to be sure their seams won't rip out under a heavy load. Be sure

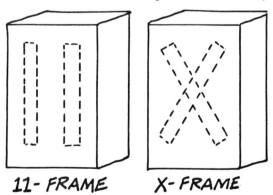

11- FRAME   X- FRAME

Example 3-4: Internal pack frame construction.

stitches and extra material reinforce the stress points where the frame contacts the fabric on all internal frame packs you are considering buying.

External Frame Backpacks

External frame backpacks can generally carry heavier and larger loads than internal frame packs because their rigid metal frames take much of the strain off the pack's less durable fabric and provide a host of places to attach extra gear when the main bag is full. An external frame design allows air to circulate between the pack and your back for ventilation and provides a rigid support for the load on your back. The external frame does, however, snag on bushes and sway back and forth as you walk because the pack rides as a separate unit on your back instead of fitting snugly and comfortably to it like an internal frame pack does.

Most frames are made of aluminum alloy tubes, while a few are made of slightly lighter but more expensive magnesium alloy. The frame is usually either an "H" or inverted "U" shape with *crossbars* added to support the main posts. Inexpensive frames have *straight ladder construction*, while better quality ones are an *S ladder design*, which is molded to the shape of a human spine. Some people claim the *hip wrap design* supports the pack's weight better on their hips, while others claim it restricts their movements and gives them a "caged in" feeling. Because they limit sideways movement and rise and fall on your hips as you walk, that style must be sized exactly for a proper and comfortable fit. *Frame extensions* are additional pieces of metal you can attach to the frame to make it larger for carrying additional equipment (Example 3-5). Test the

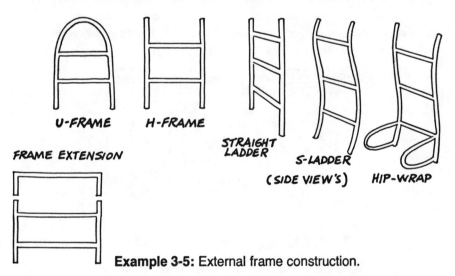

U-FRAME   H-FRAME

FRAME EXTENSION

STRAIGHT LADDER

S-LADDER

(SIDE VIEW'S)   HIP-WRAP

Example 3-5: External frame construction.

Example 3-6: Testing the strength of an external frame.

strength of a metal external frame by setting it on one leg on the floor and leaning on it. Although some models are designed to flex somewhat, don't buy one that feels soft or mushy (Example 3-6).

Packframe cross bars are connected to metal frames in several ways. *Heliarc welding,* with its uneven texture and handmade look, is considered the strongest. *Brazed welds* have a machine look, are inexpensive, and aren't completely reliable in rugged conditions. *Epoxy* is questionable. Plastic or metal *bolts, couplings,* or *screws* adjust easily, allow the frame to flex more to absorb shock, and are about as strong as heliarc welds, but aren't necessarily made with quality materials (Example 3-7).

Plastic external packframes made by Coleman Company have been on the market for a few years and appear very reliable. Other plastic frames are of questionable quality unless made by a reputable company. In general, plastic frames are significantly lighter in weight and lower in cost than their metal counterparts. They provide greater flexibility which

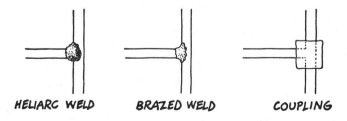

Example 3-7: Crossbar attachments to the frame.

is useful for off-trail hiking, are durable at extremely high and low temperatures, and have a host of adjusting slots to assure a proper fit.

Packbags are attached to external frames in several ways. Older styles have *sleeves* that slide over the top bars on the frame, but they tear easily at that point and must be reinforced heavily to last. Occasionally bags are *screwed* or *bolted* to the frame. *Clevis pins* in *grommets* are a popular and durable method of attachment. A small piece of cloth tape wrapped around the metal ring on clevis pins keeps them from tearing tent floors and coming apart (Example 3-8).

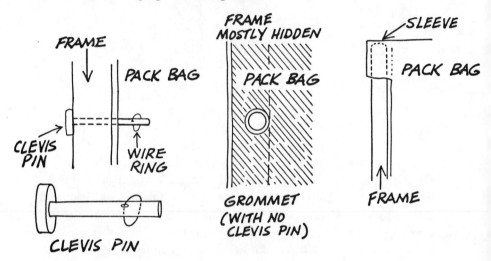

Example 3-8: Methods of attaching the packbag to the frame.

Back panels are the material that keeps the pack away from your body for ventilation and provides protection from sharp objects in it. They can be a one-piece mesh fabric or several pieces of regular or padded nylon. You should be able to tighten them and move them up and down the frame to find the most comfortable fit.

While you can buy an external frame pack large enough to fit your sleeping bag inside its packbag, most people attach their sleeping bags to the bottom of the pack frame outside the packbag. The best way to attach a sleeping bag to a packframe is with *buckled straps* which are available in camping catalogs and outdoor stores. Attaching the bag to the frame with *rope* takes longer and is difficult to do with gloved hands in cold weather, while the sharp hooks on plastic *shock cords* designed for attaching things to motorcycles snag on bushes and could tear holes in your sleeping bag.

General Backpack Characteristics

Materials

Almost all good backpacks are made of nylon, which is strong, lightweight, wind- and water-resistant, and mildew-proof. *Ripstop nylon* is sewn with extra strong reinforcing threads that stop tears, distribute stress evenly across the fabric, and allow for a lighter-weight material than otherwise would have to be used. *Taffeta* is a lightweight nylon with a softer texture and an even weave. *Ballistics cloth* is a thick, tough, almost indestructible nylon fabric. *Cordura*™ is a tough, thick material made with a very heavy nylon thread. While ripstop and taffeta are commonly used in sleeping bags and clothing, cordura and ballistics cloth are ideal for backpacks because of their durability. *Nylon pack cloth* is an abrasion- and tear-resistant material slightly less durable than Cordura.

Nylon fabric is measured by its *weight* and *thread count*. For example, 1.9-ounce nylon weighs 1.9 ounces per square yard of untreated material (waterproofing and fire-retardant coatings add weight and reduce the fabric's strength). 1.5-ounce nylon is slightly thinner and lighter than 1.9-ounce nylon, while 2.5-ounce nylon is thicker and heavier. The thread count is the number of threads per square inch of material. A fabric with a thread count of 100×120 means that it has 100 threads per inch on one side and 120 cross threads per inch on the other. A fabric with a thread count of 250 has 250 threads per inch running in each direction. The greater the thread count per inch and the larger the thread thickness is, the greater the fabric's durability, cost, and weight.

Nylon pack fabric is either *water repellent,* which means it sheds water for a short while before leaking, or treated to be *waterproof* to completely shed water. Gear in packs made with waterproof nylon still gets wet, however, because water leaks in through the zippers, seams, and other openings. Keep your belongings dry in your pack by wrapping them in individual plastic bags, lining the inside of your pack with a large plastic garbage bag, or using a *pack cover* which is a special piece of waterproof material that completely covers the pack in a storm. Using a poncho to cover a pack is not as functional because the only way to keep both yourself and the pack dry in a storm is to carry it around wherever you go, unless it's stored inside your shelter.

Stitching

Synthetic threads like nylon are more durable than cotton thread, and *double* and *triple lock stitching* is more durable than *single stitching*.

Bar tacks are extra reinforced stitching that's a sign of quality and good workmanship. Inferior quality packs have 5-6 stitches per inch, while better quality ones have 7-12 stitches per inch. Reinforced stitching and materials are needed at stress points where the pack fastens to the frame, where shoulder straps attach to the pack, and where lash cords are attached to the fabric. *Plain seams, finished plain seams,* and *serge seams* can handle a little stress, but *flat-felled seams* are much stronger. Be sure all seams are sewn away from the edge of the fabric to prevent fraying (Example 3-9).

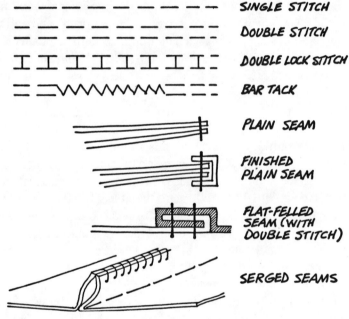

SINGLE STITCH

DOUBLE STITCH

DOUBLE LOCK STITCH

BAR TACK

PLAIN SEAM

FINISHED
PLAIN SEAM

FLAT-FELLED
SEAM (WITH
DOUBLE STITCH)

SERGED SEAMS

Example 3-9: Stitching methods.

Durability

"*Bomb-proof packs*" are typically made from tough, heavy-duty nylon materials and are designed to withstand years of rugged outdoor use. Because of their great durability, they are ideal for off-trail hiking and for hiking in areas with many rocks, cactus, or abrasive undergrowth. Their major disadvantage is that they weigh 4-6 pounds when empty. "*Tissue-paper packs*" are made from very lightweight nylon materials and are designed for weight-conscious hikers. They often weigh far less than half as much as similar bomb-proof packs. Unfortunately, their lower durability limits their use to trail hiking in mild terrain. With rough treatment bomb-proof packs provide years of service, while tissue-paper packs will last about half as long with far greater care.

Design Features

Shoulder straps, which wrap around your shoulders so you can carry the pack, should be firm and wide to best cushion the pack's weight. To insure a proper fit, internal frame packs frequently have an elaborate strap adjustment system while external frame packs have holes in their frame to adjust their shoulder straps (Example 3-10).

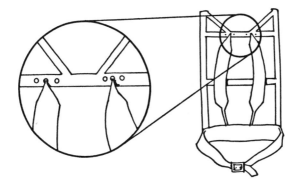

Example 3-10: Holes in external frame crossbar for shoulder strap adjustment.

A properly fastened *hip belt* transfers 60-70% of the pack's weight from your shoulders to your stronger leg muscles, because a pack should ride on your hips and not hang from your shoulders for the most comfort and the least fatigue. Hip belts should have hard (not spongy or soft) foam padding, be at least 3 inches wide, fit tight enough against your waist so that your thumb can easily slide between your shoulder and a shoulder strap, and should ride just below the top point of your hip bone. They should have a *quick release buckle* so you can unhook the belt and remove the pack quickly in an emergency.

Hip belts are either a *single belt* that wraps all the way around your waist (and acts like a padded back band against your back at the same time) or a *system of two belts* attached to the bottom corners of your frame or pack (Example 3-11). A single belt lets the pack float on your hips but tends to allow it to shift around on your back too much. The two-belt system tends to clamp the pack to your waist too tightly. Try both on with a loaded pack. If the external frame pack you want doesn't have the style of hip belt that's comfortable for you, buy a replacement belt for several dollars. Hip belts are usually not interchangeable on internal frame packs.

Quality packs have large, heavy-duty *zippers* with small pieces of fabric called *rain flaps* that cover them in a storm. Nylon or plastic zippers won't freeze shut in cold weather as easily as metal ones do. *Coil*

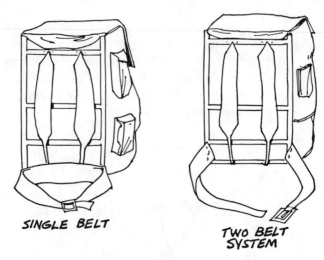

SINGLE BELT

TWO BELT
SYSTEM

Example 3-11: Hip belt styles.

zippers are just two nylon coils attached to the pack fabric. They are more dependable than standard *toothed zippers* when the zipper curves around an arc when in use.

A *sternum strap*, common on many newer internal frame packs, reduces the pressure on your shoulders and hips and stabilizes the pack's weight down the center of your body so it won't sway as much as you hike (Example 3-12).

Lash patches are pieces of leather or plastic used for attaching tents, sleeping bags, crampons, and extra gear to the pack. An *ice axe loop* is designed to hold an ice axe. *Ski pockets* are slots between the main pack compartment and its side pockets. They are designed to carry cross-country skis. All are available on common internal frame packs.

Compression straps on internal frame packs hold the packbag to-

Example 3-12: Sternum strap.

gether to lessen the strain on its zippers and prevent the contents from shifting inside the pack as you walk.

The *storm flap* should be large enough to completely cover the pack's main opening when it's fully loaded to keep your gear inside the pack and rain and dirt out.

Adjustable frames on both internal and external framed packs have movable crossbars and telescoping vertical posts so you can make the frame larger or smaller. They're useful for rapidly growing children, a family that must share their camping equipment, or for people who want to use the same pack for both mild weekend and energetic expedition hikes.

Large external *pockets* are convenient for storing often-used items like a canteen, while small pockets are the pack's "junk drawer" that hold the things easily lost. Zippered pockets hold your belongings in better than pockets closed with snaps or buckles. In theory detachable pockets increase the pack's versatility, because you can remove them when they aren't needed to save weight. Pockets are not needed if you prefer the stuff-sack method of packing your gear. It is described on page 58.

Buying a Pack

There are a number of factors to consider when buying a pack:

1) *Packbag size.* Some people recommend buying the biggest packbag you can find that's suitable for the kind of hiking you'll do because they think it's better to carry too large a pack most of the time than to have a pack adequate for many hikes but too small for longer trips. Other people prefer a smaller pack because its limited space forces them to leave any really unneeded things at home and thus reduces the weight of their pack. They feel you'll have a tendency to fill up a large pack with unneeded items.

Consider owning a small day pack for short walks and a large framed backpack for overnight hikes. Packs with up to 2,000 cubic inches of space are designed for day hikes, those with between 2,000 and 3,500 cubic inches are made for hikes up to four or five days in length, and packs larger than 4,000 cubic inches are designed for extensive backpacking and winter camping. Since those figures assume the sleeping bag is stored outside the packbag, increase them by about 1,500 cubic inches if you'll carry your sleeping bag inside a pack.

2) *Frame design.* Generally, a pack with no frame is designed for day hikes, while a framed pack is made for overnight trips. While an internal frame pack is well suited for bushwacking off trails, cross-

country skiing, scrambling, whitewater boating, and hitchhiking as well as trail hiking, external-framed packs are cumbersome and awkward for anything but trail hiking. Thus, if you need a versatile pack for somewhat shorter trips, consider buying one with an internal frame. Although not quite as comfortable as internal frame packs, external frame ones are better suited for long distance hiking because they can hold considerably more gear than internal frame models.

3) *Bag design. Single compartment* packbags are ideal for carrying very large items but it's harder to find what you're looking for in them. *Compartmentalized bags* are divided into two or more main compartments that help you organize things in the pack and regulate the pack's weight distribution better, but they limit the size of the things you can carry. For example, a watermelon would easily fit into a single compartment bag but not into a compartmentalized one (Example 3-13).

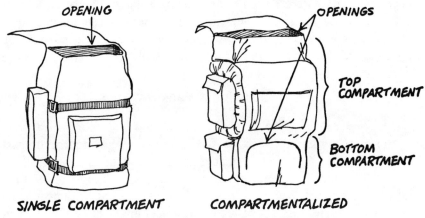

Example 3-13: Packbag designs.

4) *Packbag opening style. Top loading backpacks* open at the top of the pack, while *front loading packs* have zippered openings across the front of them (Example 3-14). You can conveniently pack top loading packs when they're standing up, but have to empty them out to get something deep inside them. On the other hand, front loading packs provide quick, easy access to anything in them but you have to lay them on the ground and pack them like a suitcase. In foul weather, you might have to lay a front loading pack down on wet or muddy ground to get something inside it.

5) *Quality and cost.* Inexpensive packs are frequently lower in cost because of inferior workmanship, poor design, or cheaper materials. If you're planning extensive backpacking trips, don't hesitate to buy the

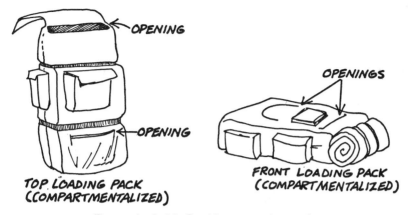

Example 3-14: Packbag opening styles.

best quality pack available, but if you only want a pack for a few relaxing weekend outings a year, you could be happy with a less expensive, lower quality model. Generally, though, a better quality pack, although costing a great deal more at first, will last many times longer than an inexpensive discount brand. The guarantee that accompanies a pack is an indication of its quality.

6) There's a great difference between a *large size* and a *large capacity*. Don't buy a large size pack if you are a small-framed person seeking a pack with a large carrying capacity.

Fitting a Pack

Finding a properly sized backpack for you is not very difficult, especially since many newer packs have suspension systems designed to fit almost anyone, and even standard framed packs typically only come in small, medium, and large sizes. Almost always, those sizes refer to frame sizes and not to packbag sizes, though. If you require a medium frame size, you might be stuck with the medium packbag that accompanies it, even though you would have liked a large bag size.

After you've selected a pack model and a size you think will fit, check that the distance between the shoulder strap attachment at the top and the waist belt attachment on the bottom of the pack is about the same as the distance between your shoulders and hips (Example 3-15). Then, with a loaded pack on your back, check that the shoulder straps are horizontal at their attachment to the top of the frame. If the strap attachments are more than 1–1½ inches above the horizontal, see if a smaller size will fit better. If they are more than 1 inch lower than the horizontal, try on a larger size.

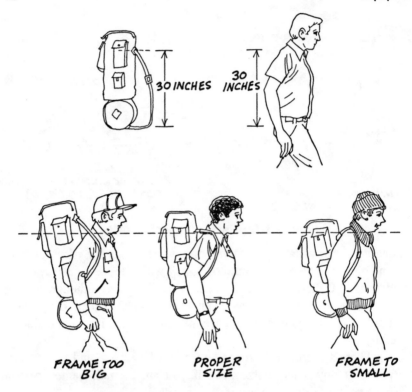

Example 3-15: Determining a proper backpack size.

When fitting an internal frame pack, note that on many models the stiff internal supports are flexible. Simply bend them as needed to assure a perfect individual fit.

Always walk around in the store with a loaded pack on your back for 5–10 minutes to be sure it fits properly. "Loaded" means the pack should contain 20–50 pounds of weights, depending on the pack's design load. Weights commonly used in stores include bags of gravel and wood blocks. Avoid heavy, concentrated weights like those used for weight lifting if possible, since they do not duplicate the weight distribution found in a backpack loaded for a hike. If you order a backpack from a mail-order catalog, be sure it fits perfectly inside your house before using it on the trail.

Putting a Pack on Your Back

Three methods describing how to put on a pack are explained below:

Method 1. Stand the pack up on a rock, a log, or your truck bumper at about waist height and slip your arms into both shoulder straps at the same time. Now simply tighten the straps and walk away from the support holding the pack. Having a friend hold it up while you slip your hands through the straps is a variation of this method.

Method 2. Pick up the pack by its shoulder straps and rest it on your right knee for a fraction of a second. Then, in one extended motion, hoist it in the air and slip your right arm through the right shoulder strap. Now, with the pack hanging from your right shoulder, put your hands underneath it, lift it up slightly, and quickly slip your left arm into the other shoulder strap. This method is easier than it sounds but considerably harder than the previous one.

Method 3. Stand the pack up on the ground, sit with your back against it and arms through the shoulder straps, roll over onto your knees, and stand up. This is the most difficult of the three methods.

After the pack is on your back, squeeze your shoulders so it lifts up a little and re-tighten the shoulder straps. Then tighten the hip belt and loosen the shoulder straps enough so that your shoulders carry about one-third to one-fourth of the pack's weight. If the straps can easily slip off your shoulders, however, they're too loose.

To remove the pack, carefully repeat any of those three methods in opposite order. Never simply drop a loaded backpack onto the ground.

Pack Care

Wrap your food, the stove, and other messy items in leakproof bags to protect your pack while hiking. At home don't store food in a pack because it absorbs odors that attract animals. See page 259 for important precautions when storing food in your pack while camping. You can wash a pack with mild soap and water occasionally when dirty, but that's seldom necessary. Clean metal zippers with rubbing alcohol and an old toothbrush when dirty (seldom if ever). Then lubricate it with a silicone lubricant. Only clean nylon zippers with an old toothbrush and soap and water. Never lubricate a nylon zipper.

Repairing a Broken Zipper

If a metal, nylon, or plastic zipper on any of your camping gear comes apart after it is zipped shut (see Example 3-16), you can fix it in a few seconds with a pair of pliers! Simply, cautiously, and gently use the pliers to tighten the outer edge of each long side of the zipper as shown in the illustration. It's better to apply a small force several times than to accidentally squeeze too hard and break the zipper. The zipper should slide noticeably tighter in its track afterwards.

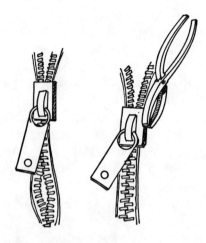

Example 3-16: Repairing a broken zipper.

4

Sleeping Bags

Outer Materials

A sleeping bag's outer material is designed to hold the insulating filler in place and shed wind, dirt, and water. It should be breathable to allow your body's water vapor to escape. Avoid waterproof sleeping bag materials occasionally found on inexpensive bags which trap evaporated body moisture that condenses and soaks the insulation.

Cotton and *polyester* are inexpensive and easily cleaned materials, but because they are heavy, readily soak up water, and are less durable than nylon, they're used primarily in lower quality bags and ones designed for car camping. Some quality sleeping bags have an "inner" outer material (the outer material that's inside the sleeping bag against your skin) made of a 50/25/25 polyester, cotton, and nylon blend. It has a slightly warmer, more comfortable feeling than 100% nylon.

Most backpacking sleeping bags have a *nylon* shell because it's a strong, durable, and lightweight material. 1.9-ounce and heavier nylon is called *downproof* because no down (see below) can slip between its tightly woven threads. 1.5-ounce nylon is not downproof and is less popular than 1.9-ounce nylon but slightly lighter in weight. Some bags are made with taffeta instead of ripstop nylon. Taffeta nylon is wind- and water-resistant and more durable and downproof than ripstop, though slightly heavier.

Some sleeping bags have a *polytetra-fluorethylene (PTFE)* outer material. PTFE materials are marketed under brand names such as

Goretex™ and Klimate™ and are discussed in detail on page 62. They are useful when sleeping with no shelter since they are windproof and completely shed external water while allowing evaporated water vapor from your body to escape. While designed to keep rain, snow, and dew away from your sleeping bag insulation and to provide an extra protective layer around it under extremely wet conditions as in the Cascades, or in suddenly wet conditions like the arid southwest, these materials don't replace a waterproof shelter.

Filler Materials

A sleeping bag doesn't generate any heat. It just prevents your body heat from escaping faster than you can produce it by forming an envelope of trapped, warm air around you. The *filler material* is the insulating material in a sleeping bag that traps air, and the thickness of that trapped air is called the *loft*. The more loft a bag has, the more it can insulate you from the cold (Example 4-1).

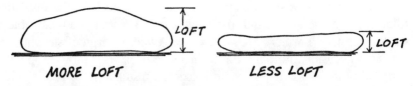

Example 4-1: Comparison of loft in cross sections of unrolled sleeping bags laying flat on the ground.

Sleeping bags contain several kinds of fillers.

Down is the warmest sleeping bag filler per unit of weight that's available and is considered the best insulating material for cold weather and lightweight backpacking trips. It endures thousands of compression and expansion cycles without wearing out (when you pack and unpack it), quickly expands when unpacked, compresses readily into a small stuff sack for carrying, and lets body moisture pass through quickly. Unfortunately, down is very expensive, difficult to clean, causes allergic reactions in some people, is useless when wet, and compacts so well that it provides no insulation under you when you sleep.

Down is the small, soft, insulating feathers on waterfowl like ducks and geese (Example 4-2). The best down comes from the most mature birds from the coldest climates, but since that down is very rare, it's quite expensive. Now most down used in camping gear comes from immature

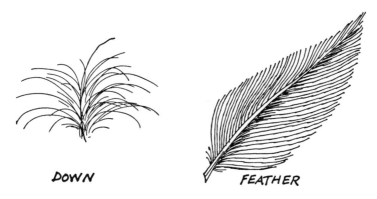

Example 4-2: Comparison of down and a feather.

birds raised in China or Taiwan. Some people claim that white down is warmer than grey down and that goose down is better than duck down, but good down is good down no matter what color it is or what kind of bird it came from. The quality of down varies considerably, though, depending on the bird's age, the composition of actual down to feathers (see below), and the manufacturers' and processors' reputations.

The quality of the down itself is the single biggest factor affecting its insulation properties. Because of high processing costs needed to get 100% pure down, the very best commercial "down" is approximately 90% down and 10% useless feathers, dirt, and scrap materials. The poorest quality "down" frequently contains more than 80% feathers and scrap materials. Down with a large percentage of scraps and feathers is cheaper, easier to process, and not nearly as warm (per unit of weight) as top quality down with an acceptable 10% feather and scrap content.

While rare and expensive down has a loft of 700–800 cubic inches per ounce (1 ounce of down will fill a container that size), good quality down from reliable manufacturers lofts to at least 550 cubic inches. However, because of variables like humidity, how the down is fluffed before measuring, and the shape of the measuring container used, it's difficult to accurately compare the loft ratings on sleeping bags from different companies. If one manufacturer claims he uses 550 down, don't assume it's better than 525 down used by another company (Example 4-3). You'll usually get what you pay for if you buy down products from a well-known company with a good reputation.

Primarily because of down's high cost and poor insulation properties when wet, several *new synthetic materials* have appeared on the market recently. *Polarguard*™ is a very stable filler material because its fibers

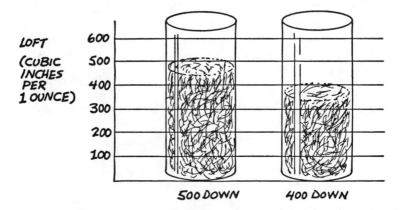

Example 4-3: Comparing loft.

are over 100 inches long. It needs no special panels and stitching to hold it in place. *Fiberfill*™ is similar to polarguard but has 2-inch long fibers which require much more stitching to stabilize them between the outer materials. *Hollofil*™ has hollow fibers that trap air inside of them for insulation. *Quallofil*™ is a lightweight material with four hollow tubes inside each fiber (Example 4-4).

Because a hollofil, polarguard, fiberfill, or quallofil sleeping bag loses only 10% of its loft and retains about 80% of its insulation value when wet, they're ideal for canoe trips and hiking in wet weather. If your new synthetic bag gets soaked miles from civilization, just wring it out and use it. Unlike down, you can wash and dry these synthetics regularly in machines according to the specific directions on their labels, they

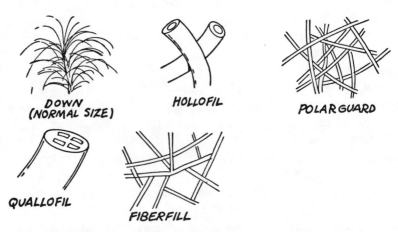

Example 4-4: Sleeping bag and clothing filler materials (magnified several hundred times).

won't give you an allergic reaction if you're sensitive to foreign substances, they won't mildew like natural fabrics do if stored damp, and they provide more insulation under you when sleeping (Example 4-5). Unfortunately they don't compress as easily and require a larger storage stuff sack or pack compartment than down does. In addition, for the same weight, they provide only about 75% of the insulation of down. A 5-pound, new synthetic bag will insulate your body as well as a down bag weighing less than 4 pounds. While quality down products last for a very long time, the life of these new synthetics is questionable, and many well-known manufacturers only market their new synthetic products with a three-year guarantee.

A host of other *old synthetic materials* like *acetate, acrylic, Kodel™, Fortrel™,* and *Dacron 88™* are less expensive but much poorer insulators than down or the newer synthetics. Bags containing these materials are designed primarily for summer car camping and indoor slumber parties. Avoid them if you plan to do any serious lightweight or cold weather backpacking. An 8-pound acrylic sleeping bag will keep you about as warm as a 3-pound quality down bag.

Several manufacturers are making *combination bags* that use two or more filler materials. Down is generally used on top because of its light weight, while a new synthetic material like Polarguard is usually used on the bottom because it offers more insulation under a sleeping person's weight.

Sleeping bags made of an *open cell foam* material are inexpensive, nonallergenic, easily washed, breathable, and warm in cold weather, but are bulky, very difficult to pack, and soak up water like a sponge. Because foam keeps its shape instead of hugging a sleeping person, cold drafts can blow up and down your body inside a foam bag. If you don't mind those problems, though, you can make your own sleeping bag (and a host of other gear like gloves, boots, and jackets) from ½- to 2-inch foam sheets available at most hardware stores for a fraction of the cost of expensive down items (see Foam pads, page 60).

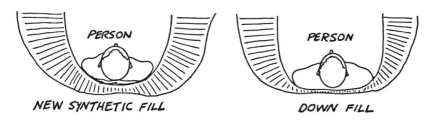

NEW SYNTHETIC FILL   DOWN FILL

Example 4-5: Compression of the sleeping bag filler under a sleeping person.

Construction

Shape

Rectangular bags provide plenty of room for people who toss and turn in their sleep and can be opened flat for use as a quilt. However they are inefficient insulators because a lot of warmth escapes through their large top opening and they have a lot of excess insulating material your body has to keep warm at night and carry around all day. They're useful when car camping in warm weather but not for backpacking.

At the other end of the spectrum, a *mummy bag* (Example 4-6) is the lightest, warmest, and most efficient sleeping bag design available. It provides the greatest insulation possible with the least amount of material and weight. Because its insulation hugs your body, there are fewer inside air spaces your body must heat and fewer drafts that blow inside the bag. Usually a drawstring turns the bag's head opening into a hood that seals in your warmth. A good quality mummy bag with 2½ pounds of 550 loft down or 3½ pounds of a new synthetic fill is adequate for almost all cold weather hikes in America. Because they're confining, however, it's difficult to dress or undress in them (useful in cold weather) and almost impossible to sleep in any position except laying flat on your back like a straight log.

A *barrel bag* (modified mummy) is basically a mummy bag with a bulge in its middle so you can dress inside easier and can curl up when sleeping. It's designed for people who feel confined inside a mummy bag but don't want to give up its efficiency.

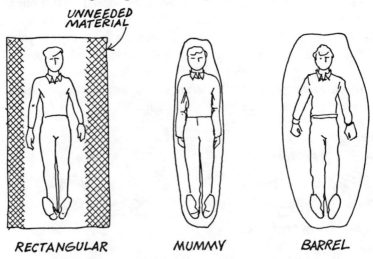

Example 4-6: Sleeping bag shapes.

Stabilizing the Filler

To be an effective insulator, the filler material inside sleeping bags must be localized in a specific position and properly secured to the bag's outer shell. This is especially important in bags with a down filler, since down clumps together and shifts around readily. *Down bags* are constructed in four major ways (Example 4-7):

1) The most inexpensive sleeping bags and ones designed for warm weather use have *sewn-through construction*. The outer material holds the inner filler material in place with stitches sewn completely through the sleeping bag. Obviously a great deal of heat can escape through the stitched areas where the insulation is compressed the most.

2) *Laminated construction* is really two sewn-through layers attached together. Few bags are constructed this way because the extra baffle (see below) and shell material adds weight and restricts the down's loft.

3) *Slant tube construction* offers the most loft for the least amount of baffle and filler weight. It's common in top-quality down sleeping bags and in bags designed for cold weather camping.

4) *Overlapping tube construction* holds the down very securely in place but its manufacturing expenses and added baffle weight are both slightly greater (several dollars and several ounces per bag) than for slant tube construction. While somewhat less efficient than slant tube construction, this method is still far more efficient than the first two methods described above.

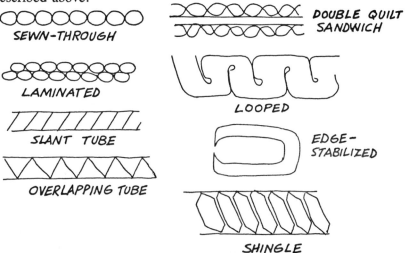

Example 4-7: Sleeping bag construction.

Good-quality down bags have thin pieces of nylon material called *baffles* sewn inside them to trap the down in compartments to get the most loft with the least amount of filler weight. These compartments should run around the bag and not from head to foot to keep the down from collecting at the foot of the bag. *Contoured baffles* are shaped to provide more loft at the top and less along the sides of a bag. They could keep you warmer because of the greater loft above you or they could give you cold spots along the sides of the bag. Be wary.

A *channel block* is a baffle used to keep the down from shifting from the top to the bottom of the bag. While most quality bags have this feature, a few purposefully don't to give you greater control of their warmth. In warm weather you shake the bag so the down shifts to the bottom where more of it gets compressed under you when you sleep and less insulation remains above you. In cold weather, you shake the down to the top for more insulation over you where it can loft to its greatest potential.

Because most *synthetic fillers* are one continuous mat of interwoven material, baffles aren't needed to prevent them from shifting throughout the bag but they must still be attached to the outer material. This is done in four ways:

1) Inexpensive and warm weather bags have *sewn-through construction*.

2) The fill in better-quality synthetic bags is held in place only along the edge near the zipper. This *edge-stabilized construction* costs more than sewn-through construction but doesn't restrict the fill's loft as much.

3) Some bags designed for colder weather combine sewn-through and edge-stabilized construction into what's called *double-quilt sandwich construction* to eliminate sewn-through cold spots.

4) The warmest synthetic bags have *shingle* or *looped construction*. The filler is layered or looped back and forth to eliminate cold spots and provide a very thick, unrestricted insulation layer.

Cut

The inner and outer shells enclosing the sleeping bag's insulation are cut in two ways. With a *space filler cut* both shells have the same circumference. The theory is that the inner shell will fold around you when you're sleeping and fill in any air pockets there. In a bag with a *differential cut*, the inner shell has a smaller circumference than the outer one. The theory here is that it's harder for you to press the inner shell against the

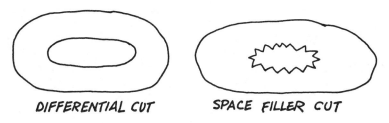

Example 4-8: Cross-sections of sleeping bag cuts.

outer shell and cause a cold spot while sleeping. Although many quality bags have differential cuts, the differences between the two are undoubtedly minor (Example 4-8).

Temperature Ratings and Loft

Manufacturers often claim that a bag is warm down to a certain temperature, but these *temperature ratings* are given only as a general guide for comparison among their own bags and not for comparing various brands. Brand X's 5°F sleeping bag is not necessarily warmer than Brand Y's 15°F bag. How much insulation a sleeping bag provides depends on a host of variables including the bag's construction, the air temperature, humidity, wind speed, location (in a tent, under a tarp, or in the open), the amount and kind of insulation under the bag, the amount of clothing worn inside the bag, how tired you are, and when and what you ate last. Manufacturers' temperature ratings are based on ideal conditions and are lower than what you should expect outdoors. For example, a bag rated at 5°F will probably keep you warm down to 10–25°F when used in typical camping situations. Generally the best indications of how warm a bag will be overall are its method of filler stabilization, kind of filler, shape, and free loft.

To measure a bag's *free loft,* unroll it, shake it gently, and lay it flat on the floor. Then just look at how "puffy" it is. The more it puffs up when empty in the store, the warmer it will be outdoors. As a rough guide, 4 inches of *total free loft* will keep you comfortably warm to about 40°F, 6 inches to about 20°F, and 8 inches to about 0°F. Since the bottom half of the bag compresses under you when you sleep, though, a more accurate way to measure its loft is to unzip it and measure its free loft for the top half of the bag only. As a general rule, you'll need 2 inches of insulation in the top half of the bag to keep warm at 40°F, 3 inches at 20°F, and 4 inches at 0°F (Example 4-9).

Because insulating material compresses under you when you sleep, many of the better bags with a channel block have a *differential fill,* which

Example 4-9: Estimating sleeping bag loft (cross sections).

means that about 60% of its insulation is on the top and 40% is on the bottom part of the bag. Thus, measuring the free loft of the top half of a differentially filled bag will give you a more accurate idea of its insulating value than measuring its total free loft (Example 4-10).

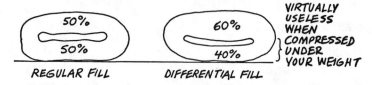

Example 4-10: Comparison of a regularly filled and differentially filled sleeping bag (cross section).

Zippers

Almost all sleeping bags designed for general backpacking use have full-length side zippers so you can adjust the ventilation for summer as well as winter camping. The better ones have reversible double zippers so you can unzip the bag from the top or bottom for heat and ventilation regulation at either end. For example, if your shoulders are warm, you open the upper zipper and if your feet are warm you unzip the lower one. Bags with side zippers can be zipped together if their zippers match, but because there's a large opening that can't be sealed when two bags are zipped together, they won't keep two people as warm as if they slept in their separate bags.

The best sleeping bag zippers are nylon or plastic because they're poor conductors of heat, seldom freeze shut, and rarely rip fabric when jammed. Oversize zippers are easier to use and more dependable than small ones. Some quality bags have *two zippers* instead of a *draft tube* (which is just an extra piece of insulation) to better prevent heat loss through the zipper area (Example 4-11). A double zipper also offers some protection if the main one breaks or snags in cold weather. Inside and outside cloth pull tabs are convenient for locating and using the zipper in the dark.

A *zipper stiffener* is a strip of stiff fabric sewn along the length of the zipper. It keeps the zipper from snagging on the sleeping bag shell.

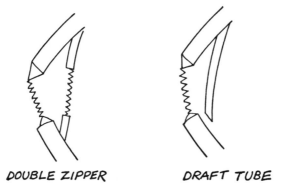

DOUBLE ZIPPER   DRAFT TUBE

Example 4-11: Sleeping bag openings (cross section).

Optional Features

A *shoulder collar,* found primarily on bags made for cold weather camping, keeps drafts from entering the sleeping bag at your neck. If your bag doesn't have one, just sleep with a shirt or sweater placed around your shoulders at night.

An *elephant's foot* is a half-length lightweight bag designed for use with a down coat or as a full sleeping bag for small children.

While *head roofs* on sleeping bags look appealing in a store, they're useless outdoors and are an indication of an inferior bag.

Some sleeping bags have a *pillow pocket* which is just a slot in which you stuff soft clothing to form a pillow. Soft clothing placed under the upper end of your sleeping bag is equally comfortable.

Drawstring hood closures are common on mummy sleeping bags. These hoods should fit comfortably and completely around your head and over your face without interfering with your breathing. When snug, only a small opening should remain above your nose. Avoid any that force you to breathe inside your sleeping bag, since moisture in your breath could condense inside your insulation. Be sure a hood fits equally well with and without a hat on your head, and be sure you can easily operate the drawstring from inside the bag.

Sleeping Bag Combinations

If you camp in areas with temperature and precipitation extremes, consider using an integrated *system of sleeping bags* to meet your needs. Instead of owning a lightweight bag for summer, a heavily insulated bag for winter use, and a synthetic bag for stormy conditions, try using the following system. Own a 2-pound down bag for generally dry weather in spring, summer, and fall. Then buy a slightly larger new synthetic bag that fits around the outside of that down bag. Use this bag alone in very warm weather, and use it with the down bag under cold or wet conditions. The synthetic material will give you that added margin of safety that's nice to have in wet weather, and both bags used one inside the other will keep you just as warm as an expedition-style winter sleeping bag. With this system, you'll gain versatility and reduce the costs of your sleeping gear.

Buying a Sleeping Bag

First, decide what kind of sleeping bag you need. Generally down is best for cold weather and lightweight backpacking trips, inexpensive polyester is suitable only for warm weather car camping, and the new synthetics are ideal in wet, damp conditions. Pick a bag that will serve your needs on most of your trips most of the time. Estimate the coldest temperatures you'll camp in, and then buy a bag designed for temperatures a little warmer than that. On very cold trips, you can always carry extra clothing to guarantee enough insulation. If you camp extensively or in rugged conditions, buy the best quality bag you can afford, because a good bag will last for years and could save your life on a cold night.

After finding a model that best suits your needs, simply try it on. Take off your shoes, crawl inside, close its zipper, and draw the hood tight around your head. Now, if you can easily scratch your back with your hand and if your toes can almost but not quite touch the bottom of the bag when your legs are straight, the bag is the proper size. It's better to buy a bag a little too big than one a little too small, because your body will press against the insulation and form cold spots in a bag that's too small. However, you should buy the smallest bag that you can comfortably fit into to avoid the excess material you'll have to keep warm and carry around, with a bag that's too large.

Judging a Bag's Quality

Generally you get what you pay for when buying from a reliable source. In addition to the information presented in this chapter, several specific ways to evaluate a bag's quality are explained below:

1) Because sleeping bag seams don't have to bear the great stress typical of backpack seams, plain seams are as satisfactory as flat-felled and bar-tacked seams and lower a bag's cost substantially. Good sleeping bags, however, have 8–12 stitches per inch of material throughout and are double-stitched at critical stress points.

2) When buying a down bag, avoid terms like "prime down," "northern down," and "100% down." A low priced bag with "down filler" usually contains a much higher percentage of useless feathers and scrap materials than an expensive garment with a well-known brand. Down insulation should feel soft and puffy. If you feel any feathers, sharp points, or lumps in a down bag when you squeeze it, avoid it.

3) If you notice your breath going through a sleeping bag when blowing into it, its material is not downproof or wind resistant.

4) With one hand inside and one outside the bag, see if you can easily push your hands together and get the inside and outside shells to touch each other with no insulation between them. If so, a cold spot will form there outdoors.

5) Understand the difference between a sleeping bag's *fill weight* and its *total weight*. A 2-pound sleeping bag is not the same thing as a sleeping bag containing 2 pounds of fill.

Sleeping Bag Care

Air out your sleeping bag between trips and every few days on longer trips to remove odors, prevent mildew, and keep its filler from matting with dirt and moisture. Always use a ground sheet (see page 58) to protect the bag from rocks and sticks, and don't use an unprotected bag as a cushion on a rock or log. Guard against sparks and fire. One spark can melt an ugly hole in a nylon shell or set the whole bag on fire. Never roll up a down or new synthetic bag. Always stuff it into a stuff sack foot end first so air in it can escape for easier packing. Store your sleeping bag unrolled hanging in a closet or laying flat on the floor, and wash it every two or three years for "infrequent use" and every year for "frequent use," according to the directions described below. Although you can wash synthetic bags far more often than down ones, washing a bag too much is worse than not washing it enough.

Cleaning a Sleeping Bag

Because it's easy to destroy a *down* bag by washing it improperly, many people recommend never cleaning one, but down absorbs body oils which attract dirt and eventually loses some of its loft and natural water repellency unless cleaned. When in doubt about cleaning a down bag, though, remember that it's better not to clean it than to clean it too much.

Dry cleaning is often the simplest method of cleaning a down bag, but not all dry cleaners know how to clean down, and harmful chemicals often linger in the bag after cleaning. Air out your bag for at least a week after it has been dry cleaned and before you sleep in it, since people have died from sleeping in bags with dry cleaning chemicals not completely evaporated out. Chlorinated hydrocarbons (Perchlorethylene) used by many dry cleaners destroys down, so be sure your bag is dry-cleaned with a mild petroleum-based compound. Since all dry cleaning solvents harm the nylon's water repellency and the down's natural oils, don't excessively dry-clean a down sleeping bag.

While some people recommend *machine washing* and drying down products, *hand washing* is safest if done correctly. Follow the directions listed below or those on a package of special *down cleaning soap* you'll need to use when washing any down garment including sleeping bags:

1) Gently press the bag into a bathtub of soapy (special down cleaning soap only!) water. Begin with the foot end so air can escape out of the top opening. When it's completely wet, let the bag soak in the water for an hour.

2) Then drain the water from the tub and use your hands to gently press as much soapy water from the bag as possible. Don't wring it out or pick it up because the weight of the wet down could destroy the bag's inner baffles and leave you with a useless bag of wet feathers.

3) Refill the tub with fresh warm water and gently press the bag to remove the soapy water from it.

4) Repeat steps 2) and 3) as often as necessary until the water running from the bag is completely free of soap.

5) Remove as much water from the bag as possible by pressing on it with your hands. Then let it rest in the tub for a day or so to let the remaining water in it drain out. Gently pat and ruffle it to help the down fluff up as it dries.

6) When most of the water has drained off, carefully lift it out of the tub by holding your hands under it to support all its weight, lay it flat on a floor or in the sun to dry, and pat it every few hours to help the down dry completely and evenly. Complete drying may take three to four days.

When camping, drying a down bag by a fire is a long and dangerous affair only recommended in an emergency.

You can wash an *old or new synthetic* sleeping bag in warm water with mild soap by hand or in a large front loading machine, and can safely dry it on low heat in a tumble dryer. Throw a pair of tennis shoes in with the bag to help shake up its filler material as it dries. When camping, you can dry a synthetic bag near a fire a lot easier, faster, and safer than a down bag. Never dry-clean a new synthetic sleeping bag, though.

Sleeping bags with a PTFE shell should be hand-washed with mild soap according to the preceding directions.

Other Sleeping Gear

Sleeping Bag Liner

A thin nylon, cotton, or polyester sleeping bag liner adds a few degrees of extra warmth, a few more ounces of weight, and years of additional life to a sleeping bag by keeping it clean. Just remove and wash the liner occasionally. Sleeping in clean pajamas or long underwear accomplishes the same thing.

Vapor Barrier

A vapor barrier is a waterproof plastic or coated nylon bag designed for use inside your regular sleeping bag in below freezing temperatures. When used properly, it lowers the effective warmth of a sleeping bag by 15–25°F. In addition, vapor barriers have several other benefits. They lessen the occurrence of dehydration in cold weather, they extend the life of your sleeping bag by keeping body moisture and oils away from its insulation, and they help keep your sleeping bag dry in damp, cold conditions. To fully understand how vapor barriers work, you need to understand the purposes of your body's sweat glands.

Your skin's pores give off water in the form of sweat to cool it when you become excessively warm. While you sweat only when overheated, those pores also emit water vapor to keep the skin's surface slightly moist all the time. At high altitudes, in areas of low humidity, or in cold weather, they can emit large amounts of water vapor just trying to keep the skin's surface moist. By sleeping in a vapor barrier, you raise the humidity of the air near your skin so it remains moist, and when that happens (and as long as you don't overheat which causes sweating), your pores stop emitting water vapor. As a result, you sleep warmer because you have less evaporative heat lost from your body and less conductive heat lost from that water vapor collecting in your sleeping bag insulation.

Because of the high humidity inside a vapor barrier, wear as few clothes inside one as possible. If you don't like the idea of touching the vapor barrier with your bare skin, wear a layer of polypropylene clothing inside it. For maximum effectiveness, be sure the vapor barrier is completely sealed around your neck or upper torso. If you sweat inside a vapor barrier, your body is too hot. Either open the sleeping bag, remove some of your clothing, or open the vapor barrier slightly.

The vapor barrier concept is also used in cold weather clothing. Vapor barrier shirts, pants, socks, and gloves are available for daytime use.

Radiant Heat Barrier

A *radiant heat barrier* (Texolite™) is an aluminized polyethylene material built into many newer synthetic sleeping bags and clothing. It reflects almost 100% of your radiated heat back to your body for an additional 15–20°F in warmth. It is also a partial vapor barrier since it greatly reduces evaporative heat loss. A sleeping bag with 6 ounces of a radiant heat barrier will keep you as warm as an identical bag with 10–14 additional ounces of fill instead of the RHB. If you have an older sleeping bag, you can buy a radiant heat barrier liner and simply insert it in the bag whenever needed. Radiant heat barriers should be washed in liquid Woolite™, rinsed twice to remove the soapy film, and dried at a moderately warm temperature.

Sleeping Bag Covers

A *sleeping bag cover* adds about 20°F of warmth to a sleeping bag, protects it from wind, external water, and abrasions, and increases your sleeping gear flexibility since it's all you need to sleep in on hot weather trips. Sleeping bag covers are usually made of polytetra-fluorethylene (PTFE) materials which are discussed on pages 43 and 62. A bivouac sack discussed on page 67 is nothing more than a heavy-duty sleeping bag cover.

Stuff Sacks

Use waterproof nylon *stuff sacks* to store and protect your sleeping bag and other gear including stove, tarp, tent, and down clothes. Stuff sacks help keep your sleeping bag dry in a storm, keep gasoline from your stove away from your clothing and food, pack dirty or wet clothes, and store food in trees away from animals. The best ones have a weather flap and a drawstring sealing their opening shut and a toggle closure on the drawstring for easier use, especially in cold weather.

Ground Sheet

A *ground sheet* is a piece of material you place under your sleeping bag and shelter floor to protect them from moisture and tears. It's a lot easier and cheaper to replace a ground sheet than to replace a torn sleeping bag or tent floor. Most of the time a sheet of 4 to 6-mil plastic that's inexpensive, lightweight, and available from hardware stores is satisfactory, although some people use special nylon ones which are more expensive but more durable. Others use their poncho or tarp for a ground sheet when weight is a critical factor on a long hike. Avoid painters' dropcloths and plastic garbage bags which are too fragile for outdoor use.

Ground Insulation

Because your sleeping bag compresses and provides very little insulation or padding under you when you sleep, you need something to insulate and pad you from the often cold and hard ground. There are three ways to achieve *comfort* and *insulation* under your sleeping bag:

Air mattresses provide excellent cushioning but offer little insulation, because the air inside of them forms convection currents which conduct body heat away. Their other disadvantages include practical jokers who let your air out in the middle of the night, greater weight, and the possibility of valve failures and punctures. Some of the newer, foam-filled designs provide a great deal more insulation and combine the warmth of a foam pad with the comfort of an air mattress. Also, you can greatly increase an air mattress's insulation value by cutting a hole in it, stuffing some down (available in outdoor stores or from catalogs) in it, sealing the hole with a patch, and putting a fine wire mesh screen over the nozzle to keep the down inside.

Air mattresses come in various styles and sizes. For a backpacking trip, use a hip-length and not a full-length one to save weight and a nylon-coated one because waterproof canvas is heavy and plastic is fragile. Consider buying a mattress that has separate air compartments instead of one large one. Although you'll have more valves to open and close, a major leak won't ruin the mattress and you can blow the outer tubes up a little harder to keep from rolling off at night (Example 4-12).

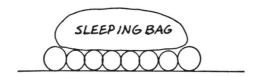

Example 4-12: An air mattress cross section showing the outer tubes inflated more than the inner ones to form a cradle to hold the sleeping person in.

Blow an air mattress up only until you feel your butt just barely touch the ground when laying on it. This makes it harder to roll off, provides maximum padding, and contains less air to carry your body warmth away. Roll your air mattress up instead of folding it to avoid weakening its material at the folds. Always carry patches in your repair kit when you carry an air mattress.

Foam pads provide more insulation but less cushioning than an air mattress. There are two kinds:

A *closed-cell foam pad* is a waterproof piece of foam with tiny, sealed-in air pockets that compress very little under your body to provide excellent insulation when sleeping in cold weather. A ½-inch thick, closed-cell foam pad is adequate for temperatures down to about 15–25°F. Consider carrying a ⅜-inch, hip-length closed-cell pad for use on cool summer evenings and carrying that plus a ⅜-inch full-length one for winter camping. Never fold a closed-cell pad, since cold spots develop along the creases. Carry and store them rolled up. Unfortunately, while closed-cell foam provides a great deal of insulation, it offers a minimum amount of cushioning.

There are two kinds of closed-cell foam pads. Polyvinyl chloride (white Ensolite™) pads become somewhat stiff in low temperatures, deteriorate with prolonged exposure to sunlight, and are heavier but more durable. Polyethylene (blue foam) pads are slightly lighter and less expensive but tear easier.

An *open-cell foam pad* is made from the puffy foam material that pads most sofas and chairs. Because open cell foam compresses a great deal under your body weight, you'll need three times the thickness to provide the same insulation as a closed-cell pad. In other words, a ¾-inch thick open-cell pad will keep you as warm as a ¼-inch thick closed-cell pad. Open-cell pads are very comfortable to sleep on but unfortunately very bulky to pack and soak up water as eagerly as a thirsty man in a desert. In most cases, it's worth carrying them only in dry conditions and on shorter hikes when your pack room is not critical.

When camping in cold weather or if weight is critical, use all your extra clothes and empty backpack instead of or in addition to a pad or air mattress for padding and insulation.

While some people like to carry a small inflatable *pillow*, a bundle of extra clothes serves just as well with less weight.

5

Shelters

Materials

Waterproof materials are complete barriers against water, while *water repellent* materials shed water for only a short time before leaking. In other words, water repellent fabrics effectively shed a light rain but are useless in a sustained storm. With the exception of PTFE materials (see below) *breathable* fabrics allow water vapor to pass through them and therefore aren't waterproof.

Plastic shelters are lightweight and inexpensive but very fragile. Often they're designed for emergency or temporary use only. While not dependable enough in wet climates like the Appalachians, they could be very satisfactory shelters in arid regions where storms are short and infrequent. *Cotton* and *canvas* shelters are practical only for car camping because they're heavy, leak when you touch them in a storm, and rot if not dried completely before storage. *Ripstop* and *taffeta nylon* are very popular shelter fabrics because they're lightweight, durable, rot-proof, wind resistant, and water resistant materials. When specially treated, they become waterproof, but the treatment occasionally cracks or peels off with prolonged use and when that happens, it's hard to re-waterproof them again. In addition, nylon gradually decomposes in sunlight. If you plan to leave a shelter set up in the sun for a very long time, consider buying one made from a material other than nylon. For the typical backpacker who plans to use a shelter for years of "average" use, though, this is nothing to worry about. Nylon is by far the most popular shelter material available because its advantages far outweigh its disadvantages. With shelters made from a waterproof *mylar/polyester* blend you lose durability as well as some weight. That material weighs one-third less than typical nylon tent material but has half the strength. One of its main advantages is that it is not affected by the sun's ultraviolet rays.

Polytetra-fluorethylene (PTFE) materials like Gore-Tex™ and Klimate™ are composed of an expensive synthetic coating applied to nylon materials to make them waterproof yet breathable (Example 5-1a). Each square inch of Gore-Tex material contains 9 billion pores, each of which are 20,000 times smaller than a drop of water (so rain can't pass through), yet 700 times larger than a molecule of water vapor (so your body moisture in vapor form can escape). Although PTFE materials are now used in a host of outdoor gear ranging from running shoes to sleeping bags, they're used primarily in rain gear and shelters. Quality PTFE shelters are generally lighter than similar nylon ones because their single-walled construction offers the same protection as double-walled nylon shelters with less material.

While waterproofness and breathability are both very desirable characteristics in a material that provides protection from the elements,

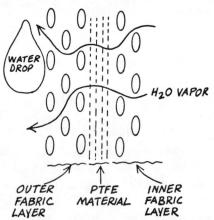

Example 5-1a: Illustration of how a PTFE material is both breathable and waterproof and of how the fragile membrane is sandwiched between two layers of fabric.

PTFE materials have some disadvantages. Some newer products contain seams sealed at the factory for waterproofness, but the seams on many PTFE garments must be hand-sealed occasionally to be completely waterproof (see seam sealing, page 75). Also, something continually rubbing or resting on a PTFE material, like a backpack's shoulder straps on a Gore-Tex parka, forces water through the pores and forms temporary microscopic leaks there. Also PTFE pores clog with dirt if not washed occasionally. These materials are not miracle fabrics. Some condensation will occur inside them under certain activity levels and weather conditions. If you sweat excessively or wear wet clothes under a PTFE outer layer, that water remains inside until it vaporizes, because water only passes through the pores in vapor form. Thus, the PTFE material could restrict the amount of water vapor that evaporates from your clothing and dead air layers under certain, uncommon conditions. In general, though, PTFE materials offer a superb combination of waterproofness, windproofness, and breathability.

Gently hand-wash PTFE clothing in mild soap and warm water occasionally, and drip or tumble-dry them on low heat. Never dry-clean them. Note, however, that shelters seldom if ever need cleaning.

Two Important Concepts

Condensation

A shelter must be waterproof to shed external water and be breathable to reduce the humidity of the air inside it to prevent that moisture from condensing and soaking your gear. Providing adequate ventilation is the best way to make a shelter breathable, since water will condense on the insides of every shelter including PTFE ones unless adequately ventilated. Condensation occurs most frequently in cold and humid conditions, and that's precisely when you need to stay the driest for safety. Example 5-1b illustrates several common sources of moisture inside a shelter. Example 5-1c illustrates the most efficient air flow pattern that naturally circulates the warmer, more moisture-laden air out of a shelter.

Foul Weather Shelters

Proper shelter is very important when camping in cold or wet conditions because of the danger of hypothermia (see page 252). In inclement weather your primary goal should be to hike safely and in comfort.

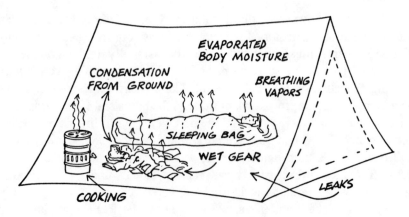

Example 5-1b: Sources of moisture inside a tent.

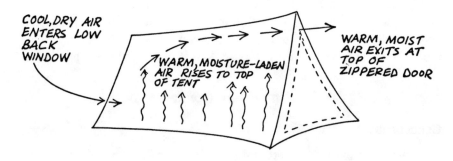

Example 5-1c: Proper air circulation that reduces condensation in a tent.

Long-distance objectives and maintaining a featherweight pack should have lower priority. The wetter and colder the weather, the more durable, waterproof, and windproof your shelter should be, and the more room you'll need inside it. The ability to sit up, turn around, and kneel is almost a necessity in inclement weather. However the ability to stand up in a shelter is a luxury whose cost is more shelter fabric and thus a heavier shelter. A typical two-person summer tent is suitable for only one person in winter because you'll want to store all your gear inside it, cook in it, and possibly even live in it for a few days if caught in a nasty storm. In typical stormy weather, you'll spend as much as 15–20 hours a day inside your shelter, so if you're buying a shelter for use in cold weather or frequently wet conditions, be sure you get one large enough to live in comfortably.

The room inside a tent is determined by two major factors. The *amount of floor area* indicates the square footage available for sleeping, storing gear, and cooking. The *amount of usable room* indicates the amount of space above the floor that you can comfortably use. A-frame tents (see below), for example, have a large floor area but a small amount of usable room inside them, since their walls slope gradually to their apex. On the other hand, dome and hoop tents (see below) provide the most usable room for a given floor size, since their walls rise steeply from the floor. A word of caution—tents always seem larger when set up in a store than when camping in them.

Tarps

Tarps are sheets of low-cost, lightweight, waterproof material. Condensation is never a problem inside them because of their superb ventilation. Limited privacy and limited wind, water, and insect protection are their major disadvantages. Tarps come in various sizes ranging from 5 by 7 feet to larger than 9 by 12 feet. Because they don't completely shelter you from blowing wind and rain, buy the largest size tarp available for use in frequently wet conditions, especially if more than one person will use it. In arid regions a very small, slightly larger than sleeping bag size could suffice.

Most tarps have metal *grommets* or pieces of cloth called *pull tabs* around their sides used to attach them to trees, rocks, or other supports. *Visclamps* (Example 5-2a) increase a tarp's versatility because you use them to attach ropes to any part of a tarp's material when setting it up instead of searching for supports conveniently lined up with its grommets or pull tabs. Simply put the rubber ball on the inside of the fabric, loop the wire around it and the fabric on the outside of the tarp, and tie a rope to

Example 5-2a: Attaching a visclamp to a tarp.

the wire's free end. You can make an improvised visclamp by bunching the tarp's fabric around a pebble and tying a rope around that to secure it. You can set up a tarp in a host of positions almost anywhere by using visclamps or this pebble-tying method, although some people carry two lightweight aluminum *poles* to set up their tarp like a pup tent in areas where no trees are handy for supports. Be careful when using visclamps with a lightweight nylon tarp. Strong winds could stretch the nylon at a visclamp attachment, ruin the waterproof coating there, and cause leaks.

Always determine the wind direction before setting up a tarp for the most protection from blowing wind, rain, and snow. Several possible positions are illustrated in Example 5-2b, but improvise others according to your situation.

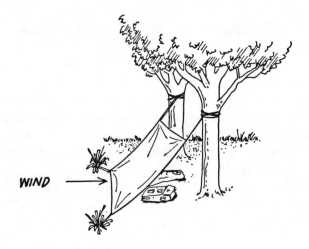

1) Pitch it like a lean-to shelter.

2) Suspend it from a rope attached to two trees.

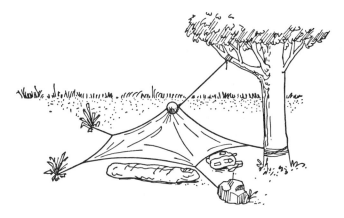

3) Secure all corners to ground supports and prop it up in the middle with a visclamp.

Example 5-2b: Different ways to set up a tarp.

Personal Shelters

Personal shelters, also known as *bivouac sacks* (Example 5-3), are small waterproof shelters only slightly larger than a sleeping bag. While some are simply a tough sleeping bag cover, others use poles to make them a small tent. In terms of cost, weight, and weather protection, they lie somewhere between the extremes of tents and tarps. Their main advantages are their light weight and compact size. They average about 2 pounds in weight and are about the dimensions of a wine bottle when packed.

Their disadvantages are many, however. Almost always, there's no room for your backpack inside a personal shelter so you'll have to leave

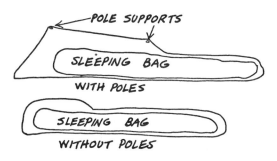

Example 5-3: Bivouac bags (cross section).

that outside in a storm. They are somewhat lonely and claustrophobic and in cold weather they are too small to cook in. While they keep you dry from a storm, moisture almost always condenses inside them in damp or cold conditions and you'll probably get wet yourself and wet the inside of one when entering or leaving one in a storm.

In general, personal shelters are useful for lightweight camping in areas with infrequent, fairly mild storms. Since ventilation is a problem with them, be sure you buy one with some kind of vent at the highest part of it and be sure there is at least one vent at each end of it to stay as dry as possible.

Hammocks

Hammocks (Example 5-4) with waterproof, tarp-like roofs and insect netting attached to them are available from some army surplus stores. They're ideal for camping in thick forests where the ground is too overgrown or too soggy to set up other shelters. The lightweight, fishnet

Example 5-4: Hammock with attached roof.

hammocks available in most stores are comfortable for sleeping but must be rigged under a tarp for adequate weather protection.

Tents

Tents offer the most privacy and insect and weather protection but cost and weigh significantly more than other kinds of shelters.

Construction

A tent must be waterproof to shed rain and snow but well-ventilated to let moisture from your body, wet clothes, and natural air humidity escape. The humidity in poorly ventilated tents often leaves a damp feeling on everything and occasionally builds up so much that a small rainstorm forms when water condensing on the inside roof drips down. Well-ventilated tents have *vents* near the roof that are sheltered from incoming rains and can be opened or closed from the inside. They also have doorway zippers that zip up and not down to a fully closed position for better ventilation of the warmer, more humid air that rises upward to the top of the tent. Some moisture condenses in tents with the best ventilation system under certain humid conditions, however.

Tents with a *single layer construction* (Example 5-5) have one waterproof barrier between the inside and the outside environments. Generally they're lighter in weight and lower in cost than tents with a double-walled construction, but almost always are poorly ventilated, less durable, and not designed for high winds, rugged use, or wet conditions. While adequate for car camping and mild backpacking trips in areas with short or infrequent storms, single-walled tents without excellent ventilation (this includes most of them) should not be used in cold weather or under severe conditions. Single-walled tents made with PTFE materials (Example 5-1a) are top quality, lightweight, waterproof yet breathable tents worth every penny of their expensive price and a glowing exception in the generally lower-quality, single-layer tent line.

Double-walled tents (Example 5-5) have one waterproof layer called

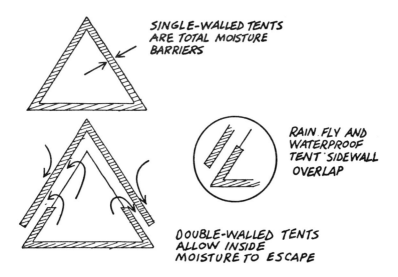

SINGLE-WALLED TENTS ARE TOTAL MOISTURE BARRIERS

RAIN FLY AND WATERPROOF TENT SIDEWALL OVERLAP

DOUBLE-WALLED TENTS ALLOW INSIDE MOISTURE TO ESCAPE

Example 5-5: Single- and double-walled tent construction.

a *rain fly* that sheds the rain and one breathable layer that lets moisture escape from inside the tent. The rain fly should completely cover doorways, windows, and vents, should be low to the ground to shed the wind, and should overlap the tent's waterproof floor. In very cold weather the rain fly adds an extra layer of restricted air which keeps double-walled tents a few degrees warmer than single-walled models.

Almost every quality tent on the market has a waterproof *floor*, which should extend 6-8 inches up its sides. A floor seals out ground moisture, crawling insects, and curious animals. If you buy a tent with a floor, you should use a *plastic ground sheet* under it for protection, since it's easier and much less expensive to replace a torn ground sheet than to replace a torn tent floor (see page 76). Except for a cookhole (discussed on page 73), there should be no seams in the floor of a tent.

Although the inside walls of most double-walled tents are made of a solid lightweight nylon material, some models are either partly or completely made from lightweight no-see-um nylon *mesh netting*. These models are slightly lighter in weight and have far more ventilation than their solid fabric counterparts. This is great in fairly warm, humid conditions, but because this extra ventilation is often excessive, the insulation value of these tents is slightly lower than those made with solid fabrics. You'll have to decide if the benefits of their added ventilation outweighs the disadvantage of their insulation loss in cold weather.

No-see-um netting is just a nylon netting similar to but much finer than mosquito netting. It is the best kind of screening material available, and should cover all tent doors, windows, and vents.

Tent Designs

Standard ridgeline tents (Example 5-6) have a straight *ridgeline* and a single pole in the front and back. They're the least expensive tent design, but the least stable in a storm. Most single-walled tents have this design.

A-frame tents are made for lightweight backpacking in all seasons and all conditions. Frequently they have a double-walled construction. The most stable A-frame tents have a *catenary cut* which means the fabric is designed on a curve to keep taut and prevent flapping in the wind. Unfortunately, A-frame tents are not designed for the most efficient use of interior space. It's difficult to sit up in them except exactly in their center.

Hoop tents are shaped like an inverted bathtub, provide the greatest

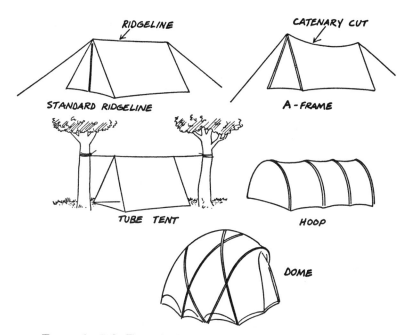

Example 5-6: Tent designs (heavy lines indicate poles).

possible interior room from an A-frame-sized floor, offer great foul weather stability, and are low in weight.

Dome tents are rounded in shape to shed snow and wind and to provide a lot more usable interior room from a given amount of floor space. Generally they're heavier, and more costly than other tent designs, but top-quality ones are popular with mountaineers on rugged expeditions. Because they're often self-supporting (they don't require stakes to hold them down), they're easy to move, clean out, and pitch in snow, but could blow away in high winds unless weighted down with backpacks or people. Most models have stakes for use in severe conditions. Their pole support system, while designed to flex in the wind and spread the stress throughout the entire frame, is often very complicated and unnecessarily elaborate.

Other tents with bizarre shapes are really slightly modified forms of the designs discussed above. They are re-engineered to try to obtain the most usable interior room with the least amount of material, the least wind resistance, the greatest ease in setting up, and the most ventilation. Some of these newer tents are very effective, while others are too complicated, too heavy, or too gimmicky.

Tube tents are simply a tube of waterproof material (usually cheap plastic or more expensive and durable nylon) open at each end that serves as a roof, ground sheet, and walls. They're inexpensive, compact in size, lightweight, and easy to set up, but unfortunately aren't very durable and leak large amounts of water at the ends if not secured carefully. Often they're good only for emergency shelters or limited solo backpacking use in areas with infrequent, mild storms. To set up a tube tent, simply hang it from a rope that's strung between two trees and crawl inside. Because they are quickly punctured by rocks and sticks, always put the same side on the ground so that the roof and sides remain completely waterproof.

Kinds of Tents

Mountain tents are small, lightweight, durable tents made to withstand rugged camping conditions. Typically they have a double-walled nylon or a single-walled PTFE construction and are A-framed, hooped, or domed in design.

Family tents are large, usually heavy tents designed for car camping or large group backpacking. They're usually domed or square in shape and are made from canvas or nylon material. Although carrying a large family tent on a backpacking trip could seem like a terrible burden, their *weight per person* (see page 74) is very low if filled to a comfortable capacity.

Discount tents are inexpensive cotton or nylon, usually single-walled, straight ridgeline tents well-suited for limited kinds of camping in mild, warm weather. They aren't designed to withstand cold temperatures, high winds, and heavy, prolonged storms because of their poor ventilation, lack of stability, and inexpensive material.

Other Tent Features

While a *zippered* door is more convenient to use for general camping, a *tunnel* entrance helps keep snow out of the tent in winter. Also, two tents with tunnels can be set up next to each other with their tunnels connected for more companionship when group camping in foul weather.

A tent's *color* is a matter of personal preference. Dark colors blend into the landscape and keep your campsite hidden from others while bright ones stand out sharply. In sunlight, darker colors let a soft, cool light in the tent, while bright colors give the appearance of being in a hot inferno (desirable in winter but not in warm weather), but bright colors add a pleasing, cheerful glow on cloudy days when dark-colored tents seem gloomy inside.

Because nylon melts very quickly and cotton burns, it's a good idea to buy a tent coated with a *fire retardant* chemical. It adds several ounces and dollars to the finished product, but slows down a fire and could be the margin of safety you need to escape when inside.

Tents designed for winter camping often have a zippered *cookhole* in their floor so spilled food falls in the snow beneath the tent rather than on the floor. In addition, during a blizzard you can retrieve snow to melt into drinking water or get rid of body wastes through it. On the other hand, a cookhole adds several ounces to the tent's weight and could develop leaks at its seams and zipper.

The best *poles* are stored side-by-side and are held together with shock cords to prevent loss and speed rigging time. *Nesting poles*—poles that fit inside each other for storage—occasionally bend and stick together, freeze shut, and jam with dirt. Fiberglass poles flex while aluminum ones are more rigid, but both are satisfactory materials. Hoop and dome tents which need flexible supports generally have fiberglass poles, while tents using immobile pole supports use aluminum poles.

Aluminum *stakes* are light but bend easily, steel stakes are stronger and less expensive but heavier, while plastic stakes are durable and lightweight but the bulkiest of all. While stakes add a few ounces of weight to your load, they'll save you considerable time each night making tent stakes at your campsite. *Snow stakes* are long, strong stakes designed for use when camping in deep snow.

In very cold weather, moisture condenses and freezes into a layer of ice crystals along the top of the tent. If you bump the tent, that ice could fall, melt, and leave your gear damp. A *frostliner* made of an absorbent, non-slick material like cotton attached to the inside roof absorbs excess moisture and collects the ice crystals. Just detach the liner and shake it out outside the tent once or twice a day.

Many tents have a *vestibule* (Example 5-7), which is a place sheltered under the rain fly or inside the tent itself where you can store your backpack and boots. It adds weight and cost but is useful in frequently stormy conditions.

Example 5-7: Snow flaps and vestibule.

Snow flaps (Example 5-7) are skirts of material attached to the bottom of a tent to secure it in snow and high winds. Simply pile snow, logs, or rocks on the flaps whenever needed, which is usually only under very severe conditions.

Pockets inside a tent are useful for storing small valuables like eyeglasses in an easily accessible, out-of-the-way place.

An interior *clothes line* running across the top of the tent is handy for "drying" (no gear will ever completely dry inside a tent) wet or damp gear in a storm and for hanging a flashlight or candle lantern.

Special features like a cookhole, vestibule, snow flaps, and tunnel entrance increase a tent's weight and cost. Only buy them if you plan to use them extensively.

Selecting a Shelter

As a general rule, a shelter's *weight per person* should not exceed 3 pounds, especially when backpacking long distances for long amounts of time. That means that a tent sleeping two people shouldn't weigh more than 6 pounds, while a tent sleeping five people shouldn't weigh more than 15 pounds. Weight is a less critical factor on shorter, less demanding hikes.

In buggy country your shelter should have no-see-um mesh *insect netting* and a sealed floor that completely encloses you, though, if you're weight-conscious, carrying a heavy shelter only for its insect protection is somewhat foolish. A can of insect repellent weighs only a few ounces while a shelter weighs several pounds.

When camping in windy areas, the shelter must offer *minimum wind resistance* and *maximum stability*.

A shelter must have enough *room* for comfort. A small shelter is satisfactory in dry climates, but you'll need a shelter large enough to store your equipment inside and allow plenty of room for moving around when camping in frequently wet conditions. The ability to sit up and turn completely around inside a shelter is extremely important in foul weather.

A shelter that's *easy to set up* is a blessing in harsh weather, while complicated shelters are nothing more than an inconvenience in pleasant climates.

A tent offers *privacy* and some protection from theft and harassment (from animals and people), while other shelters openly invite the curious and thieves.

A shelter must be properly *ventilated* to prevent a buildup of evaporated body moisture, which could condense and soak your gear. Sufficient ventilation is much more critical in cold, wet, or humid areas than in dry or breezy ones.

Buy a shelter that's appropriate for the kind of camping you'll do most of the time. If you only camp at the beach, don't buy a tent designed to withstand a mountain blizzard. Then again, don't plan on camping in the mountains with a $10 special. In terms of weather protection, if you camp in the desert or only on sunny weekends, you'll need any kind of very inexpensive shelter; if you camp in frequently wet, rainy, snowy, or windy weather, you should buy a tent; and if you need a light shelter with only modest weather protection, buy a tarp or bivouac sack.

If you decide to buy a tent, decide what general design you need— an A-frame or hoop for rugged conditions and long distance hiking, a dome for your family, or a single-walled, single-pole model for fair weather weekends. Determine which features are available for that kind of shelter and what features you really need. Do you need insect screening? A cookhole? A vestibule? A floor?

Shop around until you find a few models that have as many of those features as possible. Set them up in the store and ask yourself how much harder it would be to pitch in the wind, in the rain, or with gloved hands in the snow. Then crawl inside to see how roomy it is. When buying from a catalog, try to see the same model set up in a store first, if possible. Above all, don't try to save money when buying a shelter. A dependable shelter, like a good sleeping bag, could save your life in an emergency.

Shelter Care

Water can pass through the stitch holes around the threads, soak through the threads, or slide between the sewn layers of fabric at the seams of all waterproof materials unless properly treated (Example 5-8). As soon as you buy a shelter or raingear, fill the needle holes, cover the threads, and close the tiny gaps between the fabric layers along the seams with a seam-sealing, *waterproofing compound*, according to the specific directions on the container. Two light coats are better than one thick coat. Only seal the inside seams of clothing, since the seam sealer material leaves noticeable marks. On raingear worn only while camping and on shelters, seal both inside and outside seams for double protection. Then set up the shelter in your backyard and check for leaks in a storm or under a lawn sprinkler and re-seal the seams again if necessary. Wetting the

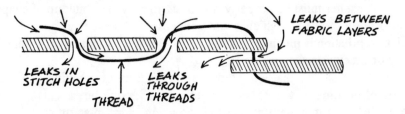

Example 5-8: Water leaking through the seams of a waterproof material.

seams with alcohol immediately before applying the seam sealing compound helps dissolve any old sealer on the seams, helps the new sealer soak into the material better, and keeps the new sealer more flexible to prevent flaking in use. Seams should be resealed at least once a year.

Spray metal zippers (on all outdoor gear) with a silicone lubricant to prevent freezing and to promote easier sliding. Attach short pieces of *shock cord* (an elastic type of cord available in hardware stores) to all guy lines to prevent torn grommets and fabric in gusty or heavy winds. Attach the shock cord as close to the fabric as possible (Example 5-9). Pack and unpack shelters carefully to prevent tears, especially if stakes and poles are stored with them. Carry stakes and poles in a separate nylon sack inside your main shelter bag to better protect the shelter fabric.

At your campsite, set up your shelter upwind and at a safe distance away from your campfire. Clear the ground of rocks and sticks before setting up a floored shelter to prevent punctures, and place a plastic ground sheet under it for additional protection. The plastic sheet should be slightly smaller than the tent floor so water can't get in and form a puddle between the sheet and the floor in a storm (Example 5-10). Keep the inside of floored shelters free of abrasive dirt, never wear boots inside, and wipe up spilled food and gasoline immediately.

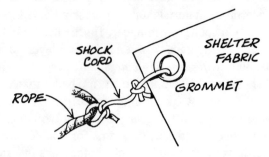

Example 5-9: Shock cord attached to a shelter grommet.

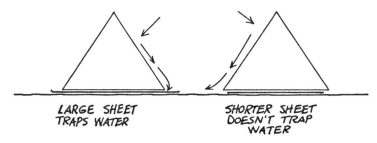

LARGE SHEET
TRAPS WATER

SHORTER SHEET
DOESN'T TRAP
WATER

Example 5-10: Prevent puddling under your tent floor by trimming the protective plastic sheet slightly smaller than the floor.

When you return from a trip, completely air out and dry your shelter, especially if it's made of cotton or if it has cotton stitching which could rot. Remove sand, twigs, and other debris from inside your shelter during and after each hike. Several times a year, vacuum tiny particles of grit out of the corners and seams along the floor to prolong the fabric's life. When dirty, sponge or wash it off with plain water or a mild soapy water solution and let it dry before storing it. Never wash a shelter with harsh detergent and never wash it excessively. Since salt water corrodes metal poles, wipe them clean with fresh water after camping near the ocean.

Occasionally high winds or sudden stress will rip a grommet out of your shelter. When you get home, simply sew a piece of waterproof nylon fabric over the damaged area for reinforcement, install another grommet with a *grommet repair kit* (available in outdoor stores), and waterproof the seams you made with seam sealer.

Some manufacturers recommend that you don't leave your nylon shelter set up in direct sunlight longer than necessary because the sun's ultraviolet rays weaken that fabric. Since the damage is so minor and occurs over a very long period of time, it's nothing to worry about when backpacking.

A Footnote About Shelters

Many popular hiking trails have *lean-to shelters* (roofed, three-sided log or stone shelters) which are often overcrowded and overused. Use an established one if convenient, but never depend on any shelter except the one you carry in your pack.

6

Clothing

Clothing has two major purposes—it *protects* you from the sun, poisonous plants, poisonous animals, bugs, and abrasions, and it *insulates* you from excessively warm or cold temperatures.

Materials

Natural fibers including cotton, wool, and silk and *synthetic fibers* such as polypropylene, nylon, polyester, and Thinsulate are the two major types of materials used in backpacking clothing.

Natural Materials

Cotton is soft and comfortable next to your skin when dry, is easy to wash, is inexpensive, and is available in almost every store. In warm weather cotton clothing is ideal. Unfortunately, it has a heavy, clammy feeling and offers no insulation when wet. Cotton clothes don't have to be saturated with water to lose their insulation value either, because even small amounts of invisible water vapor trapped by the cotton fibers drastically reduces their insulation properties. You're very susceptible to hypothermia (see page 252) when wearing wet cotton clothes in cool or cold weather, since the water in them conducts heat from your body faster than you can produce it. In hot weather, though, wet cotton clothing conducts heat from your body and helps keep you cool.

While *wool* is hot and itchy in warm weather, it's a good fabric to use in cold or wet conditions because a wet wool garment keeps you a lot warmer than a similar wet cotton garment. Since cotton has a strong wicking action that absorbs water like a sponge and spreads it throughout the fabric, you'll be wet in cotton clothes until they completely dry, while wet wool clothes quickly dry from the skin out from body heat alone. Wool clothing keeps about 80% of its insulation value when wet because it can be very wet on the outside and almost dry on the inside (Examples 6-1 and 6-2). In addition, wool garments are warmer than similar cotton ones because their naturally curly fibers resist compression and have a larger surface area to hold dead air (Example 6-3). One hundred percent virgin wool provides slightly greater warmth but at a greater price than reprocessed wool, which is made from shredded, used garments and scrap materials.

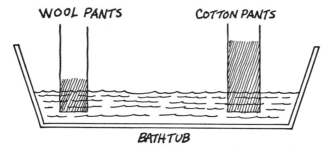

Example 6-1: A wick test showing how much more water cotton absorbs than wool.

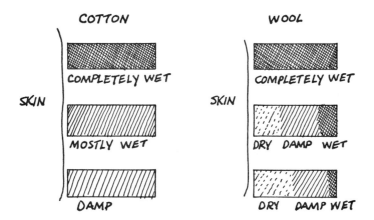

Example 6-2: Wool dries unevenly from your skin out, while cotton clothes dry evenly throughout (1-hour intervals).

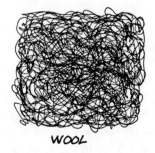

COTTON   WOOL

Example 6-3: Comparison of cotton and wool fibers.

While wool fibers bother some people's skin, top-quality soft wool irritates much less than coarse or reprocessed wool. If even soft wool bothers you, wear a very thin layer of cotton, synthetic, or a wool/cotton or wool/synthetic blend next to your skin for comfort. Remember, though, that a thin cotton undergarment chills you somewhat when wet or damp from evaporated body moisture. If wool garments don't bother your skin, remove any cotton linings in them for better insulation in cold and wet conditions.

Unfortunately, wool clothing suitable for camping is expensive. Army surplus, Salvation Army, and Goodwill thrift stores as well as yard sales and flea markets are the best places to buy wool clothing at a relatively inexpensive price.

Down (discussed on pages 44 and 45) is an excellent insulating material in cold weather garments. Its major disadvantage is that it should not be used in wet conditions since it readily absorbs water. Almost always, down garments have sewn-through construction because slant tube, overlapping tube, and laminated construction methods require a large amount of expensive hand labor and because they would be too warm for most kinds of cold weather camping. Heavy-duty down jackets should have a waist *tie string* to prevent drafts between the jacket and your other insulation layers.

Silk is becoming an increasingly popular undergarment fabric, especially since it lets moisture pass through it and feels extremely comfortable against the skin. Silk stretches readily to allow for freedom of movement, is not very bulky, and provides more warmth per weight than either wool or cotton. All silk garments should be hand-washed.

Synthetic Materials

Polypropylene is a plastic material woven into a cloth that's soft next to skin, extremely lightweight, machine washable, non-allergenic, and flexible for freedom of movement. Polypropylene allows water vapor to pass through it six times faster than through wool—and best of all, because it's plastic, it absorbs no water into its fibers at all. It stays drier in wet conditions and dries faster when wet than almost any other material used in making clothing. Because it has great water vapor-passing properties, it's an excellent material for use next to the skin in active conditions like hard hiking in cold weather and in wet conditions like storms, and is used in making hats, gloves, socks, and long underwear. Very thin polypropylene garments are designed for use under physically active conditions, while thicker (but still thin) ones are made for more sedentary activities like standing around in camp. Polypropylene should be washed fairly often to maintain its properties.

Pile is a shaggy material made from nylon, polyester, or polypropylene synthetics. Although it remains bulky when packed, is not very windproof, and pills easily (forms fiber "balls" on its outer surface), it has some qualities that make it an ideal backpacking insulating material. Pile fibers absorb only about 1% of their weight in water, so you'll feel almost as warm and dry in a soaked pile garment that's been wrung out by hand as in a completely dry one. Because of this, pile clothing is ideal in cold, damp weather. Pile provides twice the warmth of wool for the same weight, is comfortable next to the skin, is non-allergenic, and is relatively inexpensive. It is used in making insulated outer layers of clothing like jackets, sweaters, pants, and gloves.

Pile is constructed in several ways. *Fleece* is the lightest, softest, and most windproof of piles, but because it is made into a thinner material, it has the lowest loft and least insulation value. Because of this, though, it's ideal for layering clothes (see page 82). Hollow fiber pile is made from hollow fibers for more warmth with less weight. Solid fiber pile is the bulkiest and least flexible but the least expensive. Although most pile is made of inexpensive but reliable polyester, polypropylene pile is noted for its greater moisture-passing ability and lighter weight, while nylon pile is noted for its flexibility and light weight. In general, pile is pile, and differences between nylon, polyester, and polypropylene piles are insignificant compared to the differences between pile and other materials.

Thinsulate™ and *Sontique*™ (almost everywhere in this book Thinsulate is used to mean either of these fabrics) contain many fine fibers that have 20 times the surface area of down to trap a great deal of dead air. These materials are twice as warm as down although down is about one-third lighter than them for the same thickness of insulation. Unfortunately, while down easily compresses in your pack, Thinsulate and Sontique compress very little if at all. These materials are, however, still very lightweight, non-allergenic, machine washable, supple, and warm when wet. Since they absorb less than 1% of their weight in water, they are ideal for cold and wet conditions. They are commonly used as insulating filler materials in outer layers like vests, jackets, parkas, and gloves.

Radiant heat barrier and *vapor barrier* clothing are useful in very cold conditions. These materials are discussed on pages 57 and 58.

In terms of insulation when wet and comfort near the skin, *polyester* materials like orlon and acrylic lie somewhere in the middle between the extremes of cotton and wool. They don't absorb as much moisture as cotton and keep much of their insulation value when wet.

Because *nylon* is strong, lightweight, and durable, dries quickly when wet, and sheds wind and water very well, it's frequently used as an outer garment enclosing insulation layers. It's not used for insulation itself because it doesn't trap dead air as well as other materials.

Combination Materials

Many fabrics containing different amounts and combinations of wool, polyester, cotton, and nylon have the general characteristics of their base materials. *60/40 cloth*, which is composed of 60% cotton and 40% nylon, is an example. It absorbs water like cotton (though not as much as pure cotton) and is more durable than 100% cotton (though not as durable as pure nylon). 60/40 cloth is noted for its ability to shed wind and is used in making parkas.

The Layering Principle

In cold or wet conditions when the danger from hypothermia is much greater than in warm weather, you should plan your backpacking wardrobe so you can dress on the layering principle. Next to your skin wear a *vapor transmission layer*, which should include highly breathable, non-absorbent materials like polypropylene to let body moisture easily and quickly pass through. The next layer should be an *insulating layer* composed of wool, down, Thinsulate, Polarguard, or pile jackets, shirts, sweaters, pants, and the like, which trap dead air around your body for warmth. This layer is really composed of about one to five thinner layers, depending on the temperature. The final outer layer should be a *protective layer* that shelters you from water and wind.

In very cold weather, some experienced hikers recommend that you wear *vapor barrier clothing* to prevent evaporative heat lost throughout the day. They recommend wearing the vapor barrier garments either immediately next to your skin or first wearing thin polypropylene long underwear and then the vapor barrier clothes. The insulating and protective clothing layers should then be worn on the outside of the vapor barrier clothing. With this setup, you need to carefully regulate your activity level to prevent sweating. Before exercising and definitely when you begin to sweat, remove several insulating layers and/or ventilate the vapor barrier gear.

Kinds of Clothing

Short pants are comfortable when hiking in warm weather and in temperatures down to about 45°F. They can be swim trunks, cut-offs, gym shorts, or the camping shorts available in sporting goods stores, which are more expensive but somewhat more durable. While shorts offer no protection from abrasions, cold, poisonous plants and animals, and sunburn, they provide a great deal more freedom of movement and ventilation and prevent the heavy, "dragged down" feeling of long pants. One pair of shorts is all you'll ever need to carry since they wash easily and dry quickly when wet.

Generally, it's a good idea to carry at least one pair of *long pants* for warmth and protection from trail hazards on every overnight hike. Light cotton pants are not as durable but more comfortable than ever-popular denim jeans. While blue jeans or similar cotton or polyester pants are suitable for warmer weather, they are the worst thing to wear in cold or stormy conditions because they offer little insulation when wet. Wear only wool or pile long pants in cold or frequently wet weather. One pair of long pants will suffice on most hikes but carry two pair if you camp in wet or cold weather. If you carry one pair of long pants and one pair of shorts, always hike in your shorts through wet brush, in the rain, and while crossing streams, unless the weather is unbearably cold. That way your long pants will remain dry to keep you warm at night when the temperature drops. Long pants that you can put on without removing your hiking boots are especially convenient.

Use suspenders or a piece of rope as a *belt* to hold your pants up, since a leather belt interferes with your backpack's hip belt and is unnecessarily heavy and bulky. Better yet, wear pants with an elastic waistband to eliminate the need for a belt entirely.

In most places and in most kinds of weather, you'll need at least one *t-shirt* and one *long-sleeved shirt*, but carry more long-sleeved shirts if the weather is cool. If you prefer long-sleeved shirts instead of sweaters or jackets for warmth, it's better to carry several light ones than one thick, heavy one for better heat regulation.

Underwear is a personal choice. You can wear regular cotton underwear, a swim suit, or gym shorts. For overnight trips, you won't need a spare, and for extremely long hikes, you'll need two or three pairs at the most. Underwear can serve as a *bathing suit* or a bathing suit can be worn as underwear.

Cold Weather Clothing

You can wear many combinations of sweaters, sweatshirts, extra shirts, coats, and a parka to help keep warm in colder weather. Just remember that what you wear must be light, pack easily, and be based on the layering principle.

Inexpensive cotton *long underwear* is inadequate for camping because of its poor insulating properties when wet. Substantial amounts of body moisture, which conducts valuable heat away, can collect in them in less than a day of use, while an unexpected soaking in a storm can make them useless until thoroughly dried. Wool is warmer and much safer (hypothermia, page 252), but synthetic ones are the best of all. On most cold weather trips, one set of long underwear provides the extra warmth you'll need after the sun sets. If it's so cold that you'll hike as well as sleep in long underwear, carry an extra pair and use it only for sleeping to keep it as dry as possible.

A *vest* is an insulating layer designed primarily to keep the body's *core* (see page 252) from chilling. Since it insulates without restricting your arm movements, it's ideal for use when you're physically active in cold weather. Wear it over a shirt on cool days or between a parka and several insulating layers in colder weather.

A *windbreaker* is a very thin, lightweight nylon shell that reduces convective heat loss by keeping the wind away from your inner insulating layers.

A *parka* is a more durable windbreaker coat designed to keep the rain, snow, and wind away from your insulating layers as well as provide a limited amount of insulation itself. More expensive parkas made of PFTE materials protect you from wind and water, while others made of a tightly woven water repellent combination of cotton, nylon, and polyester, that leaks in anything more than an extended drizzle, offer protection only from the wind. PTFE materials offer slightly less wind protection than waterproofed, coated fabrics. Quality parkas have many pockets (though that means more seams which could leak), a hood and a high protective collar that fits snugly at the neck, a storm flap covering the zipper, wide velcro cuffs for adjusting ventilation up the sleeves, an inside drawstring closure to trap air inside your clothing better, and underarm zippers to regulate overheating. Some parkas have a specially designed snorkel hood to shield your face from the wind (Example 6-4). Shell parkas are simply a durable outer layer, while lined parkas have various amounts and types of insulation attached underneath the outer shell. Unlined, shell parkas are more versatile, especially when used with the layering principle.

A turtleneck sweater, bandana, or *scarf* keeps you noticeably warmer in cold weather by sealing your neck area from drafts and reducing the heat lost from there. Zippered turtlenecks are useful for heat regulation.

Gloves or *mittens* protect your hands in cold weather, from hot pots, and when gathering wood and scaling rocky hillsides. Mittens are warmer than gloves because your fingers can huddle together for warmth, but gloves are better where finger dexterity is required to do something, like light a stove. Consider wearing thin *glove liners* inside mittens for the warmth of mittens and protected, gloved hands when you need greater finger dexterity. When weight is critical, use a pair of socks for mittens. When camping in snow, wear a pair of coated nylon or PFTE material *overmitts* which are simply very loose fitting mittens designed to keep your inner gloves or mittens dry. While leather gloves or mittens are very durable, they soak up water and are not as warm as ones with a synthetic material like pile inside a nylon shell. Wool gloves and mittens should be worn inside an outer shell material like nylon for the most warmth.

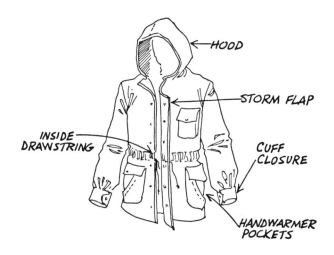

Example 6-4: Features of a quality parka.

Fingerless gloves (gloves with their fingertips open) are ideal in mildly cool weather, since they keep your hands warm yet don't limit your finger dexterity. Be careful of fingertip frostbite with them on in colder weather, though.

Depending on the style, a *hat* absorbs sweat and provides sun, rain, and insect protection. Wear a hat according to the conditions you'll encounter. When camping in the desert, carry a hat with a wide brim for shade. When camping in swampy country, wear a hat with mosquito netting attached to it. For cold weather camping, wear a ski cap, a wool stocking cap, or a *balaclava*, which has a flap that's easily changed into a face mask to keep your neck and chin warm. Consider carrying two knit hats for greater insulation in extremely cold weather.

Selecting Camping Clothing

How much and what kind of clothing you take on a hike depends upon the weather, the season, the kind of hike, your personal metabolism, and the location. Here are some suggestions to help you pack the clothing you'll need (see also Example 6-5):

1) Know the typical kinds of weather where you will hike and be prepared for the worst possible conditions there for that time of year. As a general rule, carry enough clothing to survive one night outdoors without a sleeping bag or shelter. When camping in cold weather, be especially prepared for a wet snow or cold rain that could soak your clothing and leave you with a dangerous case of hypothermia.

2) Use the same piece of clothing for different purposes. Use a bandana for a towel, pot holder, and scarf. Use socks for gloves. Use an extra shirt or pair of pants for a towel. Improvise to save weight.

3) For greater warmth, dress in layers and wear clothes that trap the most dead air. It's better to wear several thin layers than one thick one, because additional air is trapped between the layers themselves and you can regulate your body heat better than if you only wore one or two thick layers. Simply remove a layer just before exercising and add a layer just before becoming chilled.

4) Thin, loose-weave clothes are warmer and more comfortable than thick, tightly woven clothes because they allow greater freedom of movement, are lighter in weight for the same amount of trapped dead air, and allow your body moisture to pass through faster instead of lingering in the fabric and reducing its insulation value.

5) Fishnet clothes are lightweight and efficient insulators and are especially effective when covered with a wind resistant outer layer like a windbreaker, jacket, or parka that encloses the trapped dead air and

prevents convective heat loss. Fishnet clothes are warmer when wet than regular clothes because air, which remains in the spaces between the wet material, prevents conduction heat loss and hastens drying. In warmer weather, fishnet clothes are drier than regular clothes because sweat easily evaporates through the large openings in their weave instead of accumulating in the fabric, and they are cooler because of their greater convective heat loss when worn alone.

6) Wear bright orange clothing during hunting season.

7) Take two sets of clothes. Wear the same set day after day no matter how dirty they become, and save the extra set for emergencies or to keep warm when the first set becomes wet. If you have to hike in a storm and have only one dry set of clothing, wear your wet clothes on the hike and save the dry ones for the evening when you'll need the extra warmth.

8) Be sure that clothing for your upper body is long enough to cover your lower back when you bend over. If it's not, you'll have a very annoying cold spot there.

Weather	cold		warm	
Length of Hike	*2 days*	*2 weeks**	*2 days*	*2 weeks**
Clothing				
underwear	1	2-3	1	2-3
short pants	0	0	1	1
long pants	1-2	2	1	1
socks*	3-4	4-5	2-3	3-4
t-shirts	0-1	0-2	1	1-2
rain gear*	yes	yes	maybe	yes
warm clothing*	2-5	3-5	1-2	1-3

NOTES—Socks—assumes only one pair worn at a time (see page 103).

Two-week trips—assumes you wash and reuse some clothes.

Rain gear—depending on local conditions.

Warm clothing—includes any of the following: long-sleeved shirts, jackets, parkas, vest, long underwear, etc.

Example 6-5: Suggested amounts of clothing needed for various hikes (list includes all the clothes on your body as well as in your pack).

Raingear

Raingear should be waterproof to shed external water, offer freedom
of motion, be ventilated well to prevent internal condensation and over-
heating, not impair your vision, and have a minimum number of seams.
The fewer the seams and pockets, the more waterproof the garment will
be, and seams hidden in protected areas like the armpits are better than
ones exposed to the elements. Seal all seams to be sure they're waterproof
before use (see seam sealing, page 75). Buy raingear a little large to
provide more ventilation and allow room for extra insulating clothing
layers under it in cold weather.

In most cases, you must carry three different kinds of wetness
protection—a permanent shelter like a tent, tarp, or bivouac bag, a shelter
you can walk around in like a parka or poncho, and a ground sheet to keep
your sleeping bag dry when camping on wet ground. To save weight or in
dry climates, you can carry only one or two of those three kinds of
wetness protection. For example, use a poncho when hiking during the
day and as a ground sheet when sleeping at night. This is satisfactory if
you don't mind getting wet more often and if the weather is warm so you
don't have to worry about hypothermia.

Styles

There are several styles of raingear (Example 6-6).

A *poncho* is a large square sheet of material with a hole in its center
for your head. Its simple design keeps it low in price and makes it easy to
put on and remove. It can double as a ground sheet or emergency shelter,
has a lot of ventilation to minimize internal condensation, and often can
fit over your pack to keep that dry. When the wind blows, though,
ponchos flap around so much that you can't see where you're walking,
and that could be dangerous on rocky trails and steep hillsides. Their ends
snag on trees, fall into cooking pots and campfires, and catch under your
feet as you walk. They offer little protection for your arms, which get
soaked up to your shoulders, and your feet which get soaked to above your
knees. Unless the weather's very cold, it's best to wear a short-sleeved
shirt and short pants with a poncho to keep your other clothes dry for later
use.

A *cagoule* is a full-length, zippered, hooded pullover that's roomy
but doesn't billow in the wind like a poncho does. While a poncho is like
a tarp draped over you, a cagoule is similar to a large smock that encloses
you.

An *anorak* is a shortened cagoule that doesn't cover your legs. It has
a single, short zipper below the hood.

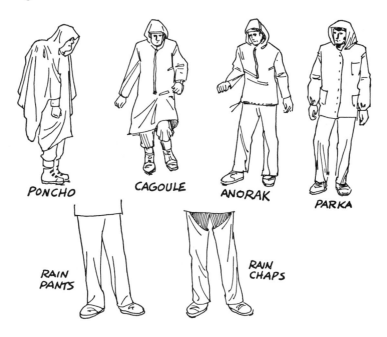

Example 6-6: Different kinds of raingear.

A *rain parka* is a coat with a full-length zipper that's shorter and fits tighter than a poncho or a cagoule. Quality rain parkas have storm flaps that keep moisture out of the zipper opening; elastic, snap, or velcro cuff closures that keep your arms dry; and snorkel hoods to shelter your head. Note that there is a difference between a rain parka and a general use parka described earlier. A rain parka is made of a waterproof material that's designed specifically to keep you dry in a storm, while a general parka is designed more to keep the wind and a very light rain from penetrating into your insulation layers.

Rain chaps are tubes of material that cover your legs but not your butt or crotch. They're lighter and more ventilated than *rain pants* which are similar to baggy dress pants in appearance and fit. Rain chaps and pants are often worn in combination with a poncho, cagoule, anorak, or parka, and should cover the tops of your boots to keep your socks dry.

Materials

As a review from page 61, waterproof garments let no liquid water pass through them, water repellent fabrics shed water for a short time before leaking, and breathable materials allow water vapor to pass through them readily. The problem with waterproof raingear is that, because it seals rain out, it prevents your body moisture from escaping. If you hike in waterproof raingear for long periods of time, you'll become very wet from your own evaporated body moisture that condensed inside it. Thus, you need a great deal of ventilation to stay dry inside waterproof raingear.

In a warm weather storm, it's best to wear a short-sleeved shirt and short pants inside your raingear or hike with no raingear on at all to keep your other clothes dry in your pack. In colder weather when you need a layer of insulation under your raingear, wear synthetics like poly-propylene and pile because they keep their insulating value when wet, and wear loose fitting clothes which allow more ventilation. Assume any clothes you wear under raingear will become damp or wet when hiking in a storm and always be sure you'll have dry clothes to wear in the evening or when the storm passes over.

Waterproof raingear is made with four kinds of materials. *Plastic* is lightweight and inexpensive but neither durable nor reliable for long backpacking trips. *Rubberized cloth* is strong and durable but unfortunately very heavy and its rubber coating often cracks or peels off after a moderate amount of use. *Coated nylon* materials offer reliable protection with little weight but significantly greater cost. PTFE materials are ideal for raingear because they completely shed water yet allow your body moisture to escape, as long as you don't sweat excessively inside them. In addition, they can double as a windbreaker or insulating parka in fair weather without causing a buildup of water vapor within your insulation layers. Materials such as No-Sweat™ and Storm-Shed™ breathe more than coated nylon fabrics but less than PTFE materials. They eventually leak in prolonged storms.

7

Footgear

Hiking footgear should provide *protection* from snakes and sharp objects; *flexibility* for comfort; *traction* in mud, snow, and wet leaves; *support* to prevent ankle injuries; *durability* for long term use; *insulation* in hot or cold weather; *ventilation* to reduce sweat buildup; and a *waterproof* barrier to shed external water. Overall, footgear should be as *lightweight* as possible while still providing the proper amount of each item underlined above for the kind of hiking you do, since 1 pound of weight on your feet equals 5 pounds on your back. In other words, walking in a pair of boots that weighs 5 pounds will slow you down as much as if you were carrying a 25-pound pack. A final characteristic of hiking footgear is *versatility*. Footgear should be functional under a variety of conditions, since conditions often change rapidly and unexpectedly outdoors. The specific kind of footgear that's best for you depends on how much of each of those factors you need for most of your hikes most of the time. There's no such thing as an ideal hiking shoe for all hiking conditions.

Old-Style Leather Boots

There are three kinds of leather boots primarily worn by backpackers. *Trail shoes* are simple boots with little or no reinforced construction. They sacrifice support and protection for their light weight. *Hiking boots* are durable, medium weight, all-purpose boots suitable for use under a wide variety of conditions. *Mountaineering boots* are designed for extreme punishment in rugged terrain. Generally, they're too heavy and stiff for most kinds of backpacking.

Construction

Leather is a very popular boot material because it's naturally water repellent yet your sweat can readily evaporate through it under most conditions. Cowhide is split into six or eight slices for making boots. The outer "fur" side of the hide is called *top grain* or *full grain* leather. Since it was on the outside of the cow, it's naturally the most water resistant and durable of the leather slices and is the best piece of leather for boots. *Suede* and *split grain* leathers are soft, porous inner slices of a hide. Because they stretch out of shape easier and are harder to keep water resistant, they're used in the construction of cheaper quality boots.

When the tougher outer side (the "fur" side) of top grain leather is on the outside of a boot, it's called *smooth out* leather. When the softer inside layer of top grain leather is on the outside of a boot, it's called *flesh out, rough out,* or *reverse* leather. While some people say that smooth out leather is better because the water resistant, tougher side is on the outside of the boot, and others claim that flesh out leather is better because the most water resistant part is protected from abrasions on the inside of the boot, the quality of the hide—top grain versus split grain—matters far more than if the top grain is rough or smooth out.

Leather undergoes a treatment called *tanning* that preserves it. *Chrome-tanned* (dry-tanned) leather is treated with chromium salts and has a hard finish, while *oil-* or *vegetable-tanned* leather is treated with natural plant materials and has a soft feel and an oily look. *Double tanning* incorporates both of those methods. All three tanning methods are similar in quality.

The *upper* is the top part of the boot (Example 7-1). Uppers made from a *single piece* of leather are strong and water resistant because they have few stitch holes which leak water or stitches that come apart, while *sectional uppers* are less expensive because they're made from odd-sized scraps of leather. Generally, the fewer the seams a boot has, the stronger, more durable, more water repellent, and more expensive it is. All upper seams should be double-stitched for maximum durability. Synthetic thread like nylon, polyester, and dacron is better than thread made from natural materials like cotton or flax because it's stronger and won't rot. Quality uppers have a narrow *backstay*, which is the strip of leather at the

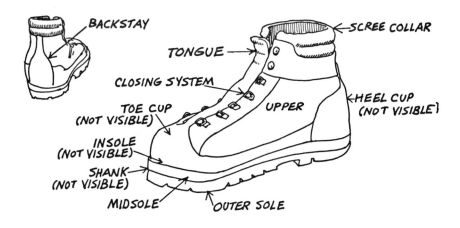

Example 7-1: The parts of a hiking boot.

back of the heel. Narrow backstays scrape on brush and rocks and tear apart far less often than wide ones.

The *sole* is the bottom part of the boot and has four parts. The *insole* is a thin piece of leather on the inside of a boot next to your foot. Often a steel, occasionally a plastic *shank* stiffens boots for more support and protects your feet from sharp rocks under them. It's more common on heavier boots than on lightweight ones. A *midsole* is an optional piece of leather between the insole and the outer sole that stiffens boots for more support, cushions your feet from sharp rocks, and insulates them from extremes in temperatures. The *outer sole* is a tough, rubberized tread that provides traction and cushions your feet from sharp rocks. A cleated *lug outer sole* provides the best traction on slippery gravel, snow, and mud, but adds weight when its cleats clog with mud. Lug soles also tear up fragile trails and delicate meadows. *Smooth-bottomed soles* are lighter in weight but not as durable overall or as dependable on slippery surfaces as lug soles.

The sole is attached to the upper in three major ways (Example 7-2):

1) In *cemented construction*, the upper is glued between and to the insole and the outer sole. Cemented boots are the lightest, most flexible, least expensive, and least durable kind of boots available. You can identify them because they have no visible stitching or midsole where the upper joins the outer sole. Cemented boots can't be resoled (see page 99).

2) *Injected molding construction* is common on boots designed for wet conditions, since the uppers are molded to a rubberized outer sole to prevent leaks.

3) Boots with a *welted construction* have their uppers stitched to the outer sole. These boots are heavier, more expensive, and more durable than the other kinds of boots and have soles that can be replaced when worn out. There are four main types:

a) *Inside-stitched* (Littleway method) boots protect the stitching inside the boot and crop the soles very close to the uppers for greater edge control. They're easily distinguished from cemented boots because they're stiffer and have a midsole. This is a very desirable construction for hiking boots.

b) In *outside-stitched construction*, the uppers are stitched directly to the midsole on the outside of the boot. Boots with this construction are relatively light and flexible, but the stitches are exposed to possible damage.

c) In *Goodyear construction*, the upper is stitched to the insole and to another strip of leather which is then stitched to the midsole. Because the uppers aren't attached directly to the soles, boots with a Goodyear welt have some degree of flexibility but aren't quite as durable as boots with other welts.

d) In the *Norwegian welt construction*, one vertical row of stitching connects the upper with the midsole and another is angled inward connecting the insole with the upper. It is the strongest, most durable method of attachment available and is common on usually heavier mountaineering and backpacking boots.

The *tongue*, the leather piece that closes the front of the boot, is designed to allow your foot to get in and out easily and to keep water out. A *gusseted* or *single tongue* is a single piece of leather that forms an unbroken barrier between your foot and the environment, while a *split tongue* has two or more overlapping pieces of leather. The bellows effect of a single tongue assures there are no gaps which could leak water but somewhat restricts access for your feet. Overall, the differences between the two kinds are minor.

Boots are closed in four different ways (Example 7-3). *Grommets* are metal circles you stick the laces through. They are time consuming to use and hard to thread with gloved hands in cold weather. *Hooks* are fast and convenient to use but poor quality ones break off or bend easily. *Swivel eyelets* are very similar to grommets. A variation of eyelets called *speed lacing* is a lot easier and faster to use than regular eyelets, because

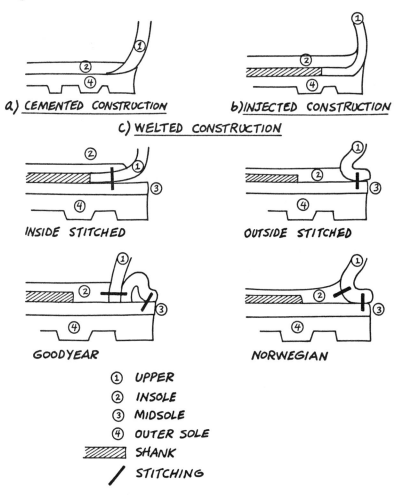

a) CEMENTED CONSTRUCTION   b) INJECTED CONSTRUCTION

c) WELTED CONSTRUCTION

INSIDE STITCHED   OUTSIDE STITCHED

GOODYEAR   NORWEGIAN

① UPPER
② INSOLE
③ MIDSOLE
④ OUTER SOLE
▨ SHANK
/ STITCHING

Example 7-2: Three major methods of sole attachment—cemented, injected, and welted. Four kinds of welted construction—inside stitched, outside stitched, Goodyear, and Norwegian.

you don't have to remove the laces from the eyelets every time you want to put on or remove your boots. *Combination lacing* systems are functional and easy to use because they incorporate eyelets or grommets on the bottom and hooks or speed eyelets on the top of the boots.

The best *laces* are either leather or nylon and not cotton or polyester. Good laces break so infrequently that you don't have to carry spares. Check them before you hike and use a piece of your emergency rope for repairs when needed in the field.

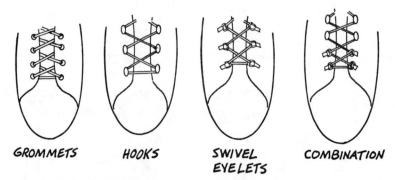

GROMMETS   HOOKS   SWIVEL   COMBINATION
                          EYELETS

Example 7-3: Boot closing systems.

A *heel cup* is a rigid piece of leather, fiber, or plastic that helps lock your heel into the boot.

A *toe cup* is a hard plastic, leather, or steel material that protects your toes from bumps.

A *scree collar* is a piece of soft leather or foam sewn around the top of the boot to keep pebbles and dirt out of the boot and for padding where the boot rubs your shin. It's effectiveness is debatable.

The more *padding* a boot has, the greater its protection, insulation, expense, and weight. Lightweight boots and boots designed for hiking on easy trails have little or no padding, while stiff boots for off-trail scrambling and boots designed for cold weather camping have a lot of padding. Wet padded boots require a long time to dry, however, and while warm in the cold, a padded boot becomes an uncomfortable sweatbox in hot weather. Buy padded and *insulated hiking boots* only if you really need their padding and insulation.

Generally, boots come either in a 6-inch ankle *height* or a 9-inch shin height. The only difference between the two is that taller boots help protect, insulate, and support your legs above the ankles, but weigh more than the low-cut variety. When selecting boots, decide if you want the greater protection of high boots or the lower weight of low-cut ones.

Buying Boots

Before buying boots, check the store's or catalog's refund policy. Most businesses will let you exchange the boots or will refund your money if you return them in new condition. When buying boots *from a store* follow these suggestions:

1) Buy boots in the afternoon or after you've walked around for a few hours to assure a better fit by compensating for your natural foot swelling when carrying a pack.

2) Wear the socks you'll hike in when trying on boots.

3) Use your regular shoe size as a starting point but select boots that are comfortable, no matter what size they are.

4) Slide your heel back as far as it goes into the heel cup of an unlaced boot and walk a few steps. If the boot falls off, it's probably too big.

5) When you're standing up in unlaced boots, you should be able to easily stick a finger down between the boot and your heel.

6) With a pair of boots laced comfortably tight:

a) Your toes should be able to wiggle freely.

b) Do several knee bends while someone holds your foot down on the floor. The ball of your foot should be snug to the sole and unable to move at all.

c) There should be no lateral movement of your foot in the boot when someone holds it firmly to the floor and you try to move your foot sideways in it.

d) A little up-and-down lift in the heel is usually unavoidable when walking in them.

e) There should be a gap of more than ½ inch between the lacing grommets, hooks, or eyelets on either side of the tongue (Example 7-4).

f) There should be a ½-inch space between your toes and the inside front of the boot so they won't jam into the boot when hiking downhill.

g) While standing on one foot, slam the other toe-first into a hard floor. Your toes shouldn't touch the front of the boot and the only feeling you should have is a snug feeling throughout your foot.

h) Wear the boots in the store for at least 30 minutes to determine how comfortable they are. Remember, they should fit snugly but still be a little stiff because they're new. Leather boots stretch slightly wider but

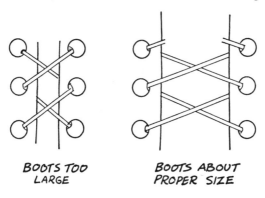

BOOTS TOO
LARGE

BOOTS ABOUT
PROPER SIZE

Example 7-4: Judging boot fit by the distance between the closure grommets.

not longer after they've been used for a while, so at first it's okay if they're a little tight in the width.

When buying boots *from a mail order catalog*, shop around until you find a style, design, and guarantee policy that fits your needs. Then, when you're ready to order, accurately measure your feet according to the following directions (and in the afternoon or after walking around for a few hours). With your weight evenly distributed on both feet, stand in your hiking socks on a piece of plain paper on a hard floor. Have a friend trace the outline of both of your feet on the paper with a pencil held straight up and down at all times. Then write your name, address, and regular shoe size on the paper and mail that in with your order.

While top-quality boots last for hundreds of hiking miles and many years, less expensive, poorer quality ones wear out much faster. Buy the best boot you can afford, but don't buy an expensive boot if it's not what you really need. There could be times when inexpensive, easily worn-out boots are more appropriate for your needs than expensive, top-quality ones. For example, it's often better to buy inexpensive boots for young children since they'll outgrow top-quality ones long before they'll wear them out.

Breaking in Boots

Wear your new boots around inside your house for a few days to give them time to soften up and conform to the shape of your feet. That way, if they don't fit properly, you can return or exchange them in new condition. After you've worn them indoors for a while and think you'll keep them, wear them outdoors on several short hikes. Finally, after you feel very confident that they fit well and won't give you any problems, waterproof them (see below) and wear them on longer hikes.

Although the procedure just described is the best way to *break in* a new pair of boots, you can quickly break them in in an emergency using the following method: First, turn on the hot water at your kitchen sink. Then quickly fill one boot up with water. Three seconds later, empty it and repeat that with the other boot. Immediately afterwards, wear them with your hiking socks until they dry. This approximates the way boots naturally soften from your sweat when you hike in them.

If boots don't fit correctly after you've worn them around your house for a while, either return them to the store or alter their fit with *heel cushions, insoles, arch supports,* or *tongue pads*. If your boots are too large, try different combinations of socks for a tighter fit. If they are too tight, some shoe repair shops can stretch their leather a little at critical pressure areas, but that should be done as a last resort.

Since *used boots* are already broken in, they're difficult to remold to fit the shape of your feet. Don't buy them if they don't feel comfortable immediately after putting them on.

Boot Care

Improperly drying a pair of wet boots will quickly destroy them. Scrape wet mud off your boots as soon as possible and while they're still wet, because mud dries out the boot's leather and the cement bonding the upper to the sole as it dries. Dry wet boots carefully and slowly by stuffing them with newspapers which gradually soak up moisture and help maintain their shape, and letting them dry at room temperature away from any external heat source. Never put your boots next to a heater, radiator, or fireplace to dry. If you must dry them by a fire, remember that if a fire feels hot on your bare hands, it's far too hot at that spot for your boots.

Applying a *waterproofing compound* softens boots, protects their leather parts from drying and cracking, and keeps your feet dry in wet weather. You should use an oil-based waterproofing compound on boots that were oil-tanned and a silicone compound on chrome-tanned ones. (Ask what kind of tanning your boots have when you buy them.) Apply the oil compound liberally, rub it in well, and allow it to dry overnight. Apply a silicone compound by holding your boots near a warm but not hot fire or stove burner and rub a chunk of sealer into them until it melts right into the leather. Many compounds soak into leather very well when treated according to the above directions or when applied and set in the sun for a few hours. Wipe any excess compound off with a cloth before a hike.

Eventually your boot soles will wear out. If your boots have cemented or injected soles, discard them, but if they have welted soles and if their uppers are still in satisfactory condition, take them to a shoe repair shop and have them *resoled*. It's like getting another pair of hiking boots at a fraction of their original cost.

New Lightweight Footgear

While the old-style leather hiking boots are durable, all-purpose boots, they have major problems. Because their leather does not breathe as fast as your feet sweat in most conditions, it traps moisture inside them. Also, since leather is not waterproof, it leaks water in wet conditions. Waterproofing the leather to seal out external water clogs its pores and seals sweat in even more. Their leather parts become stiff in cold weather and virtually useless in temperatures below 32°F when the moisture inside the leather pores freezes. They require a leather conditioner (often called waterproofing compound) to soften the leather and protect it from drying out and require a lengthy break-in time to form comfortably to your feet, but even with those things, their limited flexibility often causes blisters. Above all, they are still far too heavy for most kinds of backpacking.

The hiking footgear revolution actually began several years ago and in several different sporting arenas. As jogging mushroomed in popularity in the late 1970's, manufacturers designed innovative, lightweight running shoes based on the principle that the lighter footgear is—the farther and faster people could run. At the same time Goretex was invented. When the Goretex idea met the lightweight running shoe idea, a new and far superior kind of hiking footgear was born.

This new footgear is both reliable and extremely lightweight. A typical pair of new hiking shoes weighs about 2 pounds as compared with traditional all-leather footgear that weighs 3 to 5 pounds per pair. They are more flexible, need far less break-in time, and are much simpler to fit properly than leather boots. In addition, their nylon parts are far stronger and more durable than the leather they replaced, and when that nylon is lined with a PTFE material like Goretex, the footgear breathes yet sheds external water extremely well if constructed and cared for properly. Above all, they are extremely comfortable.

Their main disadvantage is their limited usefulness in cold or wet conditions. Your feet will stay dry and warm in them, though, if you line them with plastic bags (the vapor barrier idea, as discussed on page 57) and wear a Thinsulate-type of *overboot* (Example 7-4a) that simply wraps around the outside of your shoes for added insulation. Also, in harsh conditions, the weight you save by wearing lightweight footgear allows you to carry extra socks and a dry pair of sneakers for use around camp with no net weight penalty. If you use 2-pound nylon hiking shoes instead

Example 7-4a: Overboots wrapped around lightweight hiking shoes for extra warmth.

of 4-pound leather boots, you have in effect lightened the load on your back by 10 pounds. Although new lightweight footgear is less expensive than a pair of the old-style, all-leather boots, they are less durable when measured over a period of several hundred hiking miles.

Construction

Nylon and Goretex-coated nylon have replaced the traditionally all-leather parts of a boot's upper. Nylon is used because it will not rot or mildew and is more abrasion resistant, lightweight, and flexible than leather. However, pieces of leather are often still attached over the nylon for greater protection at critical stress areas like the toes. Boots that contain Goretex for waterproofness and breathability use it as an inner boot or liner inside the main boot to protect the fragile Goretex coating from abrasions. Some Goretex boots on the market are stitched directly through the entire boot. This negates the effect of the Goretex, since water literally pours in through the stitch holes. For the most waterproofness, the Goretex inner boot should have no exposed or untreated stitches through it at all.

Modern synthetic lightweight footgear often but not always contains a heel counter, toe cup, shank, and midsole. Generally the more of those features the footgear has and the more developed those features are, the heavier and stiffer the footgear is. These newer boots come in either a 4-inch, below-the-ankle height or a 6- to 8-inch, above-the-ankle height. Again, weight is the price you'll pay for the added support, protection, and insulation. The outer sole on many models of 1980's lightweight footgear curves around and up the front of the boot to form a tough toe protection cup. Often this begins to peel away from the upper after only a few miles of hiking. Grand Canyon rangers who prefer lightweight footgear recommend putting weatherproof epoxy glue around the top of the toe protection cup and in all the external seams (including the seam joining the upper and sole) to increase their life. This is best done before hiking outdoors with them, but should not be done on boots that can be resoled.

The uppers on new lightweight footgear are cemented, injection molded, or welted to the soles. The glue used in modern footgear is much more permanent and waterproof than that used in older-style leather boots. Whereas a few years ago, only inexpensive, easily worn-out leather boots had cemented soles, now the finest quality running shoes and lightweight hiking boots have them. (Though not every boot with cemented construction is top quality. Brand name and price are the best indicators of quality.) As with all-leather boots, the lightweight models with welted construction can be resoled. Although lightweights that can be resoled are generally slightly heavier than ones with cemented construction, they'll often last two or three times as long simply because they can be resoled.

Buying Lightweight Models

All-leather boots are difficult to try on in a store because you have to estimate how soft and comfortable they will become after you break them in. Fortunately, fitting the new lightweight footgear is much easier and less risky than fitting the outdated leather boots. Simply try them on and walk around in them for a few minutes in the store. If they fit comfortably there, they'll fit comfortably on the trail. When fitting lightweight footgear in a store, all the suggestions explained on pages 96-98 apply. When ordering them from a mail-order catalog, follow the procedure explaining how to order all-leather boots described on page 98.

Care

Usually the new lightweight models need no break-in period, although dealers still recommend that you wear new boots indoors for a few days to be sure they fit well. That way, if they don't fit properly, you can return them in new condition for an exchange or refund. Their nylon and Goretex uppers need no special waterproofing/conditioning treatment. Just wipe dirt off them with a damp cloth occasionally. The leather parts on some models, however, should occasionally be treated. Ask about this when you buy your boots. Wash nonleather footgear in mild soap and rinse well after an extremely dirty hike or several times a year to remove dirt and sweat salts from them.

Other Footwear

Insulated hiking boots have a layer of insulation sealed between two layers of leather, rubber, or nylon. While comfortable in cold weather, they're unbearably hot in warm weather and difficult to dry when wet. Buy them only for extensive cold weather use. Closed-cell foam insulation in these boots is thin and warm but reduces the amount of moisture that can evaporate out of the boot, while open-cell foam insulation is breathable but soaks water up like a sponge and takes a long time to dry.

Shoe pacs are boots with rubber soles injection molded to leather uppers and containing felt liners for insulation. They are made for hiking in wet or cold conditions. Carry extra liners and socks to guarantee dry and warm feet if you hike in them.

Mukluks are large, thick, usually foam-insulated boots designed for limited hiking in very cold weather. They're comfortable when worn around camp or after a day of hiking, snowshoeing, or skiing and light enough to carry on moderate backpacking trips.

Sneakers are suitable for warm weather backpacking under a variety of conditions. Although they wear out quickly, offer little ankle support and protection, don't insulate your feet from heat or cold, get wet easily, and cost more than quality boots in the long run, their superb ventilation and extremely light weight often far outweigh those disadvantages. Generally, boots are better for very rocky or slippery trails and for people with weak ankles, but sneakers are better for people who value light weight at the expense of support, protection, and insulation. They're also ideal for children who quickly outgrow expensive boots. While sneakers get wet very fast, they dry much quicker than boots, and you can help insulate them for use in colder weather by wearing two pairs of wool socks and using the plastic bag insulation method described on page 193.

Several companies are marketing *heavy-duty sneakers* for backpacking that typically have a canvas or a water resistant nylon upper glued or molded to a rubber or plastic lug sole. They provide more padding, protection, and ankle support than regular sneakers with only a slight increase in weight.

Regular leather *army surplus boots* and surplus *jungle boots* with a leather-reinforced nylon upper and lug sole bottom are durable, lightweight, and inexpensive.

Consider carrying light *sandals, moccasins,* or *sneakers* to wear around your campsite in the evening. The extra comfort they provide is often worth their additional weight on undemanding, relaxing backpacking trips.

Socks

Socks should *cushion* your feet, *absorb perspiration, prevent friction* between your foot and the boot, and *insulate* your feet from cold or hot ground. Generally, wool socks are ideal because they're more durable, provide greater padding, and keep your feet warmer after they've become damp from body moisture than cotton socks—although cotton ones are cooler on your feet in warm weather. Some people hike in one pair of thick wool or cotton socks, but others who are bothered by wool's itchy feeling, wear a thin nylon or cotton sock as a liner between their feet and their wool socks. Others believe that wearing two pairs of socks prevents blisters because the two socks rub against each other as they walk and release the friction between feet and hiking boots. Those people either wear two thick socks or a thin cotton or nylon liner sock under a thick outer sock. Another variation is to use one pair of socks with rubber or foam *insoles* which provide more padding and insulation under your soles. Try different combinations of socks and insoles until you find what works best for you.

When hiking, carry several pairs of socks and change them frequently to keep your feet dry, since damp socks insulate poorly and cause blisters. Always keep a pair bone dry for sleeping in at night, even if you must put wet ones on in the morning. Wash socks frequently to remove blister-causing sand and dirt particles from them.

Equipment suppliers will replace socks with a *mileage or time guarantee* if they wear out within a certain number of hiking miles or time period. Avoid all socks that are classified as long-wearing if they have no specific guarantee.

Insulated socks are designed to be worn inside your sleeping bag or tent in cold weather. Typically they have down or a synthetic insulation like Polarguard inside a thin nylon outer shell. Often they have elastic tops that hug your ankles.

Booties

Booties are specialized, insulated socks that have a tough nylon outer sole for durability and a closed-cell foam insole for insulation. They are suitable for limited walking on snow or bare ground, like when you need to leave your tent in the middle of the night to tap a kidney.

When buying booties, check for these features:

1) The quality and amount of *fill* ranges from down in expensive ones to quilted cotton in cheap ones.

2) Their *size* ranges from about 4 inches to about 8 inches in height.

3) Leather *soles* are slippery and absorb water, while nylon ones are slippery and not as durable as leather.

4) The *insoles* should be closed-cell foam for maximum insulation.

5) They have elastic or tie *closures* at the top, which are both equally effective.

Gaiters

Gaiters are pieces of waterproof nylon or PFTE material that keep snow and water out of the gap between the top of your boots and the bottom of your pants and keep the bottom of your pants dry (Example 7-5). They're most useful in cross-country skiing or winter hiking in

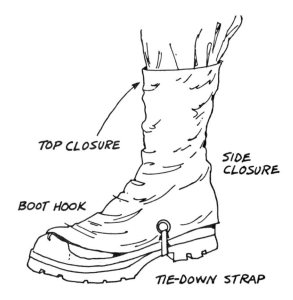

TOP CLOSURE

SIDE CLOSURE

BOOT HOOK

TIE-DOWN STRAP

Example 7-5: Gaiter.

snow, but some people use them for keeping dry when hiking in wet weather and for preventing cuts when hiking in deserts. Better gaiters have a *tie-down strap* running under the boot sole, a *boot hook* that attaches to your boot laces to keep them from sliding up your legs, and a *top closure* to keep snow from falling down through their top opening. Their elastic, velcro, snap buttons, or zipper *side closures* should be at the back of the gaiter above your heel to keep snow and water from entering as you hike. Tie-down straps made of cord often freeze and are time-consuming to remove in even the most ideal conditions. Buckle tie-down straps are easier and faster to use. If you do use cord, however, try leaving it tied permanently. Simply slip your feet into them instead of untying their knot each time you put them on or remove them.

8

Cooking Gear

Stoves

Most backpackers cook on stoves instead of wood fires because they're fast, convenient, and simple to use. With a stove you don't have to limit your camping to locations with plenty of available firewood in order to cook a hot meal. In fact, nowadays you need a stove for cooking in many popular hiking areas because they've been stripped clean of firewood from overuse. With a stove your meal can cook while you set up your campsite, your pots won't turn black from campfire soot, there's less chance that food will burn onto the pots, and you can even cook food on a stove without getting out of your sleeping bag or when stormbound inside a tent. Their primary disadvantages include their added weight and limited cooking capacity, which somewhat restricts your menu. Generally, you can only cook one pot of food at a time on a backpacking stove, so most backpackers only cook one-pot meals like soup, stew, or oatmeal.

There are three major kinds of fuel used in backpacking stoves, and stoves are generally categorized according to those types of fuel.

White Gas

White gas is a refined type of gasoline readily available in most sporting goods stores. It's inexpensive, effective in cold weather, and burns with a high heat value. Although it's potentially explosive and requires priming (see below), it's one of the most popular fuels for backpacking, especially among experienced hikers.

White gas stoves have several major components (Example 8-1). A cloth *wick* draws liquid fuel from the *fuel tank* to the *vaporizing tube* where it vaporizes, passes through a *nozzle jet* to the *burner*, and burns. A *burner plate* spreads the fuel vapor out so it can burn easier, and a *valve* controls the amount of vapor traveling through the vaporizing tube to the nozzle jet to regulate the stove's heat output.

Gas stoves must be *primed* to use. That means you must raise the temperature of the liquid fuel enough so that some of it vaporizes and can ignite. The colder the stove and fuel temperatures are, the harder the priming process is, although even under the most severe conditions, gas stoves aren't difficult to prime. Priming is more of a chore that has to be done than an obstacle to overcome. Most gas stoves are self-pressurized, which means that when lit their own heat vaporizes the liquid fuel, while a few have a *pressurizing pump* to help prime them in colder weather.

Before priming a gas stove, open and close the fuel tank cap to get rid of the vacuum that formed there as it cooled the last time it was used and to check the level of the fuel in the tank. Refill it if it's low, but not more than three-quarters full in warm weather, since too much fuel in the tank interferes with its vaporization and makes the stove harder to light.

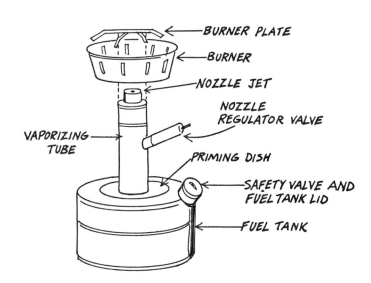

Example 8-1: Parts of a typical white gas and kerosene stove.

In cold weather, however, a stove with a completely filled fuel tank lights faster. A good way to prime a gas stove is to hold it over a wad of burning paper for 10–20 seconds or so. Another method is to remove some of the liquid gas from the fuel tank with an eyedropper, place it in the *priming dish* located above the fuel tank, and light it with a match. Then, when that priming flame goes out, turn the control valve on and light the stove's burner. As a general rule, prime a gas stove with 1 eyedropper of fuel in mild weather and 2 eyedroppers of fuel in cold weather. A yellow or sputtering burner flame means that you didn't prime the stove enough, while a blue flame indicates a properly primed stove.

If you buy a gas or kerosene stove, you'll need a pint or quart *fuel bottle* to carry extra fuel, but buy one that's specifically designed for storing fuel and not just water. A *funnel* or *fuel spout* helps you transfer fuel from the fuel bottle to the stove's fuel tank without spilling it and a plastic *eyedropper* helps you prime the stove. A *nozzle poker* cleans the nozzle jet whenever it clogs or when the stove doesn't seem to burn properly, but many newer stoves have a self-cleaning needle inside them which replaces this tool (Example 8-2).

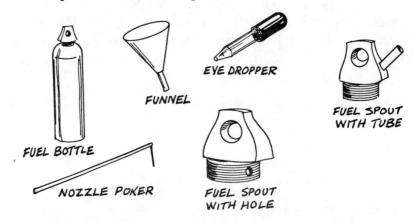

Example 8-2: Additional items needed when using a gas or kerosene stove.

Kerosene

Kerosene is inexpensive, reliable in cold weather, and efficient, but leaves greasy stains on your gear and burns with a smoky, smelly flame. Because it has a much higher vaporization point than gas, kerosene stoves are a lot safer to use but much more difficult to prime than gas ones. You must use a small amount of white gas or chemical firestarter instead of kerosene from its fuel tank to prime a kerosene stove. Kerosene stoves are similar to white gas stoves in design.

Butane

Butane fuel is popular among beginning hikers because it's extremely simple to use, easy to ignite, requires no priming, and involves no messy liquid fuel. Simply attach a canister of fuel to the stove burner, open the valve, and light it. There are several disadvantages with butane fuel, however. While butane stoves themselves are inexpensive, compact, and lightweight, butane fuel is many times more expensive and has half the heat value of gasoline or kerosene. Thus over a period of years, butane stoves cost a great deal more to operate than other stoves. Also, they require a significantly longer cooking time and more fuel to cook the same amount of food. Since butane cartridges depend on atmospheric pressure to operate, they're almost useless in temperatures below freezing unless kept warm in your coat pocket or sleeping bag prior to use. Because the canisters lose pressure as they burn, it could take up to twice as long to cook food with a used cartridge as with a new one. Because it's almost impossible to tell how much fuel remains in a used cartridge, you almost always have to carry more than you really need, and their non-disposable empty cartridges must be carried out of the wilderness and disposed of properly.

Butane stoves have a very simple design. A metal *stem* connects the *fuel canister* to the *burner*, and a *valve* controls the stove's heat by regulating the amount of vapor passing through the stem. While most butane stoves have a *vapor-feed* system where the pressurized liquid butane vaporizes automatically when the control valve is opened, *liquid-feed* models use cartridges with built-in wicks for slightly better cold weather use (Example 8-3).

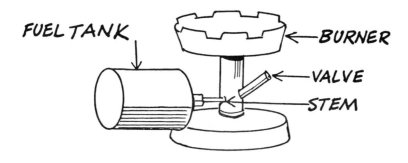

Example 8-3: Parts of a typical butane stove.

Other Fuels

   Propane fuel is reliable in all temperatures and requires no priming, but is stored in steel containers usually too heavy for backpacking use. (See also Example 8-4.)
   Alcohol stoves are seldom used for backpacking because alcohol has one-half the heat value of gas, is expensive, and is difficult to find in stores.

Fuel	Advantages	Disadvantages
white gas	efficient, high heat output inexpensive easy to obtain effective in cold temperatures stove fuel used for priming	must be primed potentially explosive
kerosene	efficient, high heat output inexpensive easy to obtain very inexplosive effective in cold temperatures	must be primed (carry gas or firestarter) smelly, smoky flame leaves greasy stains difficult to ignite spilled fuel doesn't evaporate quickly
butane	no chance of spills burns very clean low stove weight and cost simple to use (no priming)	not reliable below 32°F empty containers not disposable outdoors expensive hard to judge amount of fuel in used containers containers lose effectiveness as they burn
propane	no chance of spills effective in cold temperatures no priming required	heavy, nondisposable containers expensive
alcohol	very inexplosive	low heat output difficult to obtain expensive
solid fuels	inexpensive simple to use	empty containers not disposable outdoors low heat output

Example 8-4: Comparison of stove fuels.

Solid chemical fuels are safe, relatively inexpensive, simple to use, and require no priming, but have a terribly low heat output that makes them very unreliable in even moderately cool weather and useless in cold weather.

Safety

Backpacking stoves are very safe when used properly. Never use one in high traffic areas where it could get kicked over, or near burnable items like leaves or camping gear. When cooking in a tent, be careful you don't bump the stove and set the tent on fire. Allow plenty of ventilation when cooking in a tent so excess steam from your food escapes instead of condensing on your equipment, so the stove won't consume all the available oxygen in it, and to prevent a buildup of carbon monoxide. Never fill a gasoline stove inside a tent, because the tent could trap the vapors and explode when you light the stove.

Before lighting a *white gas* or *kerosene* stove, be sure its fuel tank lid is closed, the valve is off, and the extra fuel canister is capped and away from the stove. Most gas and kerosene stoves have a *safety valve* that releases when the pressure in the fuel tank becomes too great. It prevents the stove from exploding all at once like a bomb but turns it into a 3-foot-long flame thrower instead. Prevent explosions by never refilling or opening a stove's fuel tank until it's turned off and has cooled down. Never use automotive gasoline in a stove designed only for white gas, because impurities in it will eventually clog it. When storing a stove at home, remove the fuel from it to reduce its fire hazard and to prevent stored fuel from gumming up its insides and affecting its performance.

While propane and butane *canisters* eliminate the risks of explosions possible with an unprotected fuel like gasoline, they're equally dangerous overall. By taking the precautions mentioned above, you can virtually eliminate the dangers associated with gasoline and kerosene stoves, but canistered fuel is more of an insidious hazard, since you never know when the canister's connection to the stove will fail or the canister itself will leak (though rare, those things do occur). In warm weather, never cover a fuel canister or restrict the flow of air around it to prevent over-heating. If a canister gets hot, turn off the stove immediately.

Try a new stove in your backyard first so you can learn how to use it under controlled and safe conditions. At first, stoves seem like complicated and dangerous contraptions to people not familiar with them, but they're reasonably safe and simple to use with a little practice.

Factors Fuel—	White Gas	Kerosene	Butane	Propane	Alcohol	Solid Fuel
1) Availability of fuel	good	average	good	good	poor	average
2) Lightweight (stove, canister, and fuel) (per unit of cooked food)	good	good	average	poor	average	poor
3) Heat output (per unit of cooked food)	good	good	average	average	average	poor
4) Convenient and simple	average	average	good	good	good	good
5) Reliable in cold weather	good	good	poor	good	average	poor
6) Cost of stove	average	average	good	good	good	good
7) Cost of fuel (per unit of cooked food)	good	good	poor	average	poor	poor
8) Simmers food easily	poor	poor	good	good	good	good
9) Heats quickly	good	good	average	good	average	poor

Example 8-5: Factors to consider when buying a stove.

Odds and Ends

You can get *replacement parts* for your stove from large camping stores or catalog suppliers. See the Appendix for addresses.

The *amount of fuel* you'll need varies with the kind of stove, the types of foods you'll cook, the altitude, and the weather conditions. As a general rule, allow about 1 quart of white gas or kerosene, three butane canisters, one 32-ounce canister of propane fuel, 2 quarts of alcohol fuel, and many 4- to 6-ounce containers of solid fuel for two people per week of summer camping. Double those amounts for cold weather or high altitude camping or when purifying water by boiling for drinking. When camping in cold weather, place the stove on your foam pad to insulate it from the cold ground to conserve fuel.

Buying a Stove

The factors to consider when buying a stove are listed on the chart (Example 8-5) on page 112, but only you can determine the relative importance of those factors to your particular needs. While that chart compares stoves in terms of their general fuel characteristics, its information necessarily varies somewhat, depending on particular stove models. Other important factors like durability, size, resistance to wind, compactness, and stability vary widely depending on specific brands.

Pots, Pans, Utensils, and Things

The kind and amount of cooking gear you need depends on the *weight* you're willing to carry, the *kinds of foods* you'll eat, the *number of people* in your group, and the amount of *available room* in your pack. Generally backpacking cooking gear should be lightweight, compact, and functional and your cooking process simple. *Group gear* is equipment like the stove, soap, and cooking pans which are shared by everyone, while *individual gear* includes the pocket knife, spoon, cup, and plate each person uses themselves. For solo hikers, there is no distinction between the two categories.

When weight is not very critical, there's virtually no limit to the amount and variety of cooking gear you can carry. However, when saving weight matters more than eating a fancy meal, individual gear should consist of only a pocket knife, a spoon, and a cup. Use the cup for drinking liquids as well as for eating meals, because plates are really unnecessary. Solo hikers can eat food directly from the pan it's cooked in, while groups of people must carry at least one or two pots to cook their food in. If you really need to save weight, cook over a fire or eat food not requiring cooking and leave your stove at home.

If you're just getting started in backpacking, you can use metal *pots* and *pans* from home for a while, but eventually you'll probably buy an aluminum *cook set* consisting of a variety of pots and pans because they're lightweight, compact, and convenient for backpacking. In addition, cook sets usually *nest* (fit) together to save room in your pack. Small sets serve one or two hikers, while more elaborate ones are designed for larger groups.

Plastic *cups* and *dishes* insulate hot and cold foods, while aluminum and stainless steel ones burn your lips and hands when hot, quickly warm cold food on hot days, and cool hot foods in cold weather, but can double as cooking pots. Metal or plastic *telescoping cups* fold into a compact unit for storage. Durable plastic *spoons* sold in camping stores are lighter than typical metal silverware. Plastic butter dishes are ideal for hiking because of their light weight.

Hikers relying on campfires to cook their meals occasionally carry a small, lightweight *backpacking grill* wrapped in a plastic bag or nylon stuff sack. This is definitely a "luxury" item that lightweight hikers would never even consider carrying.

Depending on the amount of available water you'll find on your hikes, you'll need at least one and possibly up to four *water containers* to carry and store water. Try to buy containers that fit into the outside pockets of your pack so you can get to them easily when thirsty. Plastic water containers are very durable, weigh half as much, and cost considerably less than aluminum ones. Since plastic containers absorb the taste of their contents, though, always store water flavored with powdered drink mixes in the same container and carry another one which always contains pure water. Never mix powdered mixes in aluminum containers, which absorb the taste of the mix and deteriorate from chemicals in them. *Bota bags* are ideal for carrying water or alcoholic beverages, since they're lightweight, compress easily as they empty, and are easy to carry on day hikes. Their only disadvantage is they could break unexpectedly in your pack if it's handled roughly. Many hospitals discard (and will gladly give away if asked) lightweight plastic *saline solution containers*. These make ideal canteens after they've been rinsed out. When setting up a base camp or hiking in dry conditions, consider carrying a 1-gallon plastic *collapsible jug*, which is light, durable, and easily packed when empty. Never carry water in plastic *milk jugs*. They are extremely unreliable outdoors.

Knives

Sheath knives have a long, strong, immobile blade protected by and stored in a piece of leather, while *pocket knives* are short, compact, fold up inside themselves for storage, and often contain a variety of tools like a can opener and a file. While sheath knives are designed for heavy duty slicing like skinning game, pocket knives are made for general cutting and are the only kind of knife you'll need for backpacking (Example 8-6).

SHEATH KNIFE   POCKET KNIFE

Example 8-6: Kinds of knives.

A simple pocket knife with a blade, can opener, awl, and bottle cap remover is much lighter and more functional than the often overrated, multi-purpose Swiss Army knife containing tools you'll never need or use. Stainless steel knives are better than steel ones in wet climates because they won't rust.

A knife will last for years if cared for. Always store sheath knives in their holders and pocket knives folded up. Never stick them into trees, logs, fires, or the ground. Use them for cutting and slicing but not for chopping. Dry and clean all knife blades immediately after use. Clean dirt from the grooves in a pocket knife and oil its joints occasionally. Tie a loop of colored rope to it to find it easier in your pack. A sharp knife is a lot safer to use than a dull one, so keep it sharp at all times.

Since most new knives have a machine-sharpened blade, it's a good idea to refine the edge by hand-sharpening it yourself after you buy it. To *sharpen a knife* properly (Example 8-7), you need to use a *sharpening stone* suitable for that purpose and available in stores; any rock you find laying around is not satisfactory. A soft, medium-grained stone is a general-purpose sharpening stone designed to sharpen a relatively dull knife, while a hard, fine-grained stone is used for a final polishing and touching up of an already sharp blade.

1) Put a small amount of honing oil or light oil on a sharpening stone. Use water if no oil is available. Never sharpen a knife on a dry stone.

2) Lay the blade flat on the stone with its cutting edge facing away from you.

3) Tilt the back of the blade up at a 15–25° angle. The number does not matter, but maintaining a constant angle is critical.

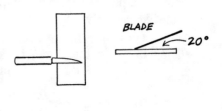

4) Push the blade away from you as if you were trying to cut a thin slice off the stone. Repeat two or three times in the same direction.

5) Turn the blade on its other side, hold it at the same 15–25° angle, and repeat step 4 by pulling it toward you. Do the same number of strokes on this side as you did on the first side of the blade.

6) Repeat this procedure, alternating sides of the blade until it's sharp. A properly sharpened blade edge will not reflect light when held edgewise in the sun, and will slice a piece of paper cleanly when drawn across its edge.

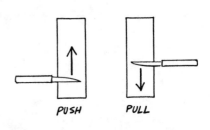

7) Rinse the stone with water or wipe it clean with a towel when finished.

Example 8-7: How to sharpen a knife.

Other Odds and Ends

Washing

You need *soap* for washing dishes, washing clothes, bathing yourself, and coating the outside of pots before cooking with them over a fire to remove campfire soot from them easily. Liquid soap in a small container could leak in your pack or spill at your campsite, while a small bar of soap is more compact, safer to carry, and requires no special care. The small motel-sized bars are ideal, but if none of those are available, simply break a regular bar in half or use leftover pieces from home. To save weight and bulk, carry a bar of soap in a plastic bag and not in a plastic soap dish. Liquid castile soap sold in health stores under a variety of brand names is ideal for backpacking because it is concentrated, biodegradable, and can be used instead of toothpaste for brushing your teeth. Use about 2 drops for brushing your teeth and about 10–15 drops for a complete sponge bath or rain shower. A small *scrub pad* (a sponge is not abrasive enough) is handy for cleaning dirty dishes. A few backpackers carry a portable plastic *wash basin* to wash their dishes and clothes in, but that's not really necessary. Your largest cooking pot or a plastic bag will suffice.

Matches

Because paper *matches* soak up water, burn out quickly, and are harder to light in cold, wet, or windy conditions, carry wooden matches on your camping trips. Strike-anywhere wooden matches are the best of all because you don't have to carry a striker to use them. Store all matches in waterproof containers or plastic bags to keep them dry, and carry them in several bundles scattered in various places in your pack for even greater protection from loss or moisture. Although relatively expensive, specialty windproof and waterproof matches are ideal for use in severe conditions when you simply have to get a fire lit. It's a good idea to carry a small bundle of them in your survival kit.

A small *cigarette lighter* is heavier but gives off a stronger, longer flame, which is useful in frequently wet conditions. If you carry a lighter, always carry matches too, because a drop of water landing on a lighter's flint could make it useless until it dries out. For emergencies, things that provide a flame, like matches or a lighter, are much more dependable than things like flint and steel which only give off sparks.

Firestarters are often useful when building a fire or lighting a stove in cold or wet weather. You can use a candle stub, dried paper wads, the gas from your stove, or the commercial firestarters available in stores.

9

Additional Equipment

First Aid Kit

For your safety, always carry a first aid kit when hiking anywhere. A homemade kit is more compact, less expensive, and better equipped than the kind sold in stores, because you can package them to meet your particular needs. While store-bought kits are usually adequate for short weekend trips, they're frequently under-supplied for solo hikers or for hikes in rugged terrain. A solo hiker could pack a homemade first aid kit in a small metal bandaid box or plastic soap dish, while a first aid kit for hiking groups, which requires greater amounts of items, could fit in a pint-sized food storage container. What you carry in it depends upon the length, nature, and location of your hike, but the minimum items every first aid kit should contain are listed below:

1) A *needle* for puncturing blisters.
2) *Tweezers* for removing splinters and cactus thorns.
3) An assortment of *adhesive bandages* for small cuts.
4) A *razor blade* for emergency cutting.
5) A small roll of *adhesive tape* for taping gauze pads over cuts. Remove the metal parts to save weight.
6) Several 2- and 4-inch *gauze pads* for larger cuts.

7) A small tube of *bacterial cream* to disinfect cuts.

8) One roll of 2-inch-wide *gauze* for larger cuts.

9) *Aspirin* sealed in plastic for headaches and body pains.

10) *Moleskin* patches for blisters.

11) *Anti-acid tablets* sealed in plastic for an upset stomach.

The following optional items are useful in certain areas:

1) A very small bottle of *ammonia* (or Benadryl tablets available with a prescription) for bee stings. Just rub the ammonia on the stung area as soon as possible to kill the pain.

2) A *snakebite* or *anti-venin kit* for poisonous snake bites. Be sure you know how to use them before hiking.

The most important item you'll need, however, and one that's not available in stores, is a working knowledge of first aid skills. Chapter 21 explains how you can deal with possible emergencies and common minor problems on the trail, but reading it or carrying a written chart accompanying a store-bought first aid kit are poor substitutes for proper first aid training. Never go far from help if you or someone in your group hasn't taken a Red Cross first aid course.

Repair Kit

A *repair kit* should contain items like self-sticking ripstop *repair tape* to fix holes in nylon materials, a *needle* and *thread, rubber bands, safety pins,* extra *buttons,* extra *plastic bags, air mattress patches,* a 5- to 20-foot piece of thin nylon *rope,* and an extra *stove valve control key.* Exactly what you carry depends on where you're hiking, how long you'll be gone, and the kind of equipment you have. Repair kits are obviously much more useful on long, remote hikes than on short weekend ones.

Survival Kit

A *survival kit* contains things you'll need if you become lost or stranded outdoors. It can be combined with the first aid kit and repair kits if you desire. Here's a minimum list of the things it should contain:

1) Several *dimes* and *quarters* and a list of *emergency telephone numbers.* This is especially important for solo hikers and people with unusual medical problems.

2) Several very large colored toy *balloons* for emergency signals.

3) Waterproof *matches*, candle stubs, and firestarters.

4) A few fishing *hooks, sinkers,* and some *line*.

5) A *whistle* to blow if lost.

6) A large, clear *plastic bag* for use as an emergency ground sheet, water collector, shelter, and solar still (see solar still, page 279).

While some people recommend carrying *emergency food* like bouillon cubes and candy bars, it makes more sense to plan your food menu so that you always have a little food left over from a trip. That way you'll eat well and still be prepared for emergencies. A bouillon cube or two won't help you much in an emergency anyway, and it's impractical to cart a case of emergency food around all the time. You should have other survival items, like a signaling mirror, compass, pocket knife, first aid kit, plastic sheet or tarp, and water purification tablets elsewhere in your pack.

Health Supplies

Use a bandana or small washcloth instead of a full-length *towel* to save weight and room in your pack. Carry a small *toothbrush*, a very small tube of *toothpaste,* and some *dental floss*. If weight is critical, forget the paste and brush your teeth with salt or plain water. Carry a *comb* and a *razor* if you desire but use *soap* for shaving cream and shampoo. That soap should be the same that you use for washing dishes and clothes to save weight and simplify your gear. Carry a small metal (not glass) *signalling mirror* that has a small hole in the middle and reflects on both sides (see page 283). Use *toilet paper* for bathroom needs, blowing your nose, cleaning dishes, and starting fires. Always protect it with plastic bags. As a general rule, two people will use about one 500-sheet roll each week. Carry a small container of *foot powder* and put some in your boots while hiking to prevent blisters in hot weather.

Because inexpensive plastic *sunglasses* reduce the visible light rays without restricting the ultraviolet rays, your pupils dilate with them on and actually absorb more dangerous ultraviolet rays than if you left them off. Use only top-quality sunglasses to prevent sun and snow blindness in bright conditions, which usually include hiking in high altitudes, in snow, and across deserts. *Glacier glasses* (Example 9-1) are specialized sunglasses with flaps that reduce the amount of reflected sunlight entering through their sides.

Example 9-1: Glacier glasses.

Carry a *sun screen* to prevent sunburn, especially at high altitudes, when hiking or camping in snow, and in continually bright conditions. A tube of cream is lighter and easier to carry than a container of liquid or spray. Carry *lip balm* to keep your lips from drying out.

Diethyl-meta toulamide (DEET) is the best chemical *insect repellent* protection available. Look for it on the list of ingredients of commercial preparations. The greater its concentration, the more effective they are. Small roll-on tubes and bottles of liquid are lighter and more compact than aerosol spray cans. Wet your skin with water before applying concentrated repellents for more effective and economical use of them. In wet conditions try mixing your insect repellent with petroleum jelly so it won't wash away.

You'll need a small bottle of *water purification tablets* or vial of liquid chlorine bleach or iodine if you hike where you're not absolutely sure the drinking water is safe. Iodine and chlorine are the two major types of tablets available. (See purifying water, p. 174.)

When hiking near populated areas, a self-defense *attack spray* gives some protection from angry dogs that could threaten you.

The Little Extras

A *walking stick* helps you maintain your walking rhythm, improves your balance when crossing streams, checks bushes for snakes along the trail, knocks rainwater and dew off plant leaves in your path, becomes a pole for rigging your shelter, and provides psychological protection from wild animals, but many people avoid them because they're a lot of unneeded weight to carry around all day.

While a large two-D-cell *flashlight* provides a strong light beam useful for hiking at night and collecting firewood after dark, a small penlight saves a lot of weight and provides all the light needed for general camp use. In addition, a flashlight small enough to hold in your mouth frees both hands for critical chores like untying knots and lighting the

stove. No matter which kind of light you decide to carry, switch one of
the batteries around inside it, tape the switch down, or secure the switch
with several rubber bands to prevent the light from accidentally switching
on inside your pack. Carry extra *batteries* on long trips and an extra *bulb*
on every trip. Try to buy a flashlight that stores an extra bulb inside it, if
possible.

Lithium batteries are the lightest, most dependable source of power
available for backpackers. While *nickel-cadmium* and *carbon/zinc* batteries dim after 1–2 hours of use and *alkaline batteries* last only slightly
longer, lithium batteries last all night and only fade immediately before
dying completely. At temperatures below zero, carbon/zinc and alkaline
batteries are about 10–15% efficient, but lithium ones are over 80%
efficient. Although lithium batteries cost a lot more than other types of
batteries, their longer, more dependable life makes them cheaper to use in
the long run.

Candles provide a pleasing, mellow light after dark. Lightweight
aluminum *candle holders* protect them from the wind and make them safe
enough to use inside a (well-ventilated) tent.

Binoculars are useful for route-finding when traveling cross-country
in steep, open terrain. A few hikers in the Rocky Mountains carry them
for sighting grizzly bears in open country.

Cord-locks (Example 9-2) are a convenient way to attach two thin
strings together. They are commonly used on sleeping bag drawstrings,
on flaps that cover a backpack, and on gaiters. You can purchase them
from most camping supply stores and can easily attach them to your gear.
Better quality ones are a one-piece construction that won't fly apart when
you remove the string that comes with them. To adjust one, simply press
on each end with your thumb and index finger and slide it along the
string.

A tiny *army surplus can opener* (Example 9-3) is handy for opening
cans when your pocket knife doesn't have a can opener blade.

A 50-foot piece of 400-pound test nylon *rope* is handy for lowering

STRING ENDS
KNOTTED TOGETHER

Example 9-2: A cord-lock.

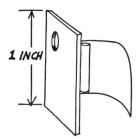

1 INCH

Example 9-3: Can opener available at army surplus stores.

packs down cliffs, hanging food in trees and rigging a clothesline. Small pieces of thinner nylon rope make handy spare bootlaces and all-purpose straps.

While some people hike with a *wristwatch* to keep track of how far they've walked in a certain amount of time or how much time they have left until dark, you'll only need one if you must be at a destination at a specific time to meet other hikers or your ride home. A watch should be small, lightweight, shock-resistant, and waterproof. An *alarm clock*, preferably on a wristwatch, is useful if you want to get up early for sunrise photographs or need to begin hiking early in the morning. Put it in a covered metal cooking pan and set that near your head at night so its loud rattle will wake you in the morning.

You should carry some kind of *identification*, especially if you hitchhike or use developed campgrounds; *fire and camping permits* in certain areas and national parks; *maps* (described on pages 210-214) and *guidebooks* when hiking on an unfamiliar trail; a *compass* (described on pages 215-219) for map reading and survival; *spare eyeglasses* and a *carrying case*, especially if you have to drive home; extra *rubber bands* for securing things; extra *plastic bags* for storing gear, waterproofing your equipment, making vapor barrier liners for your hands and feet, storing dirty or wet clothes, and carrying trash; a *pen* and *notebook* for keeping notes or leaving messages; *money* for emergencies or to buy things at stores along the way; *rainy day items* like a book to read, cards, sketch pad, or chess set; *sports equipment* like fishing rods, binoculars, and camera gear; a *day pack* if you plan to set up a base camp and day hike from it; a small plastic *trowel* for digging a latrine when group camping; and a maximum/minimum *thermometer* if you're interested in knowing the temperature. That type of thermometer is useful for testing your personal endurance limit and the usefulness of your gear at different temperatures.

Mark small or frequently used items like a pocket knife, tent stakes, and compass with fluorescent orange paint to find them easier when lost.

Unneeded Items

A *saw* or *small axe* is not needed for backpacking. If you can't collect firewood and build a fire without them (see firebuilding, page 205), use your stove or don't camp out.

A *pedometer* is an instrument used to measure how far you've walked. Since it's based on the length of your stride, it's inherently inaccurate on backpacking trips, because that varies with the differences in terrain, the weight in your pack, and how tired you are. A map is much more accurate for measuring distances.

Serious, lightweight backpackers never carry a *camera* because it's a lot of dead weight to constantly carry around.

10

Food

Selecting Food for a Hike

Nutrition and Energy

Your body uses *proteins* to build and repair tissues, *fats* for great amounts of sustained energy, *carbohydrates* for quick energy, *vitamins* and *minerals* to regulate cell chemistry and body functions, and *fiber* for good digestion. While a balanced diet contains about 25% fats, 25% proteins, and 50% carbohydrates and consists of eating a variety of foods containing a satisfactory amount of all the items listed above, you can live for a long time on foods containing an unbalanced mixture of them. Therefore, on all hikes except for very long backpacking trips and as long as you eat a reasonably varied diet, the only thing you need to consider when planning your menu is getting enough calories (energy). You don't need detailed calorie charts or a calculator for that either. Just remember that you'll eat more food while camping than you do in your regular daily lifestyle. Typically you burn up about 1,500–2,500 calories in an average day, but you need 2,000–3,000 calories a day for a summer hike, 3,000–4,000 calories a day on a hard backpacking trip, and over 5,000 calories a day when hiking in cold weather. You can take in more calories simply by eating more food or by eating foods containing greater amounts of fat, which contains twice as much energy per unit of food as proteins and carbohydrates.

Simplicity

Carry a wide selection of food if you want to relax and feast, but if you don't want to spend a great deal of time cooking, eat foods requiring little or no preparation, like instant foods that just need added water and pre-cooked foods you just heat up. Generally, backpacking foods should have uncomplicated directions and be easy to cook. Try to avoid foods requiring more than 10-15 minutes of cooking time. Also, carry foods that require no specialized packaging or care. Avoid cans which are heavy, jars which break, and refrigerated foods that spoil.

Personal Tastes

Choose foods to suit your personal tastes and eating habits. While people naturally are less picky with what they eat when camping, you'll have a more pleasant hike if you eat foods you really enjoy rather than foods that you are supposed to eat because they're "backpacking foods." It's best to avoid drastic, sudden changes in your diet, especially for short, weekend hikes. Gradually acclimate your body to "backpacking food" one or two weeks before going on a demanding or extended hike when the constraints of weight and bulk necessarily restrict what foods you carry.

Weight and Bulk

The kind and amount of food you carry depends on how far, how long, and when you'll hike. Since weight and bulk are usually not critical on short hikes and food won't spoil in a few hours, carry any kind of food you want to for day hikes and almost any kind of food on weekends, including foods requiring refrigeration (especially in cold weather or if there's a cold stream near your campsite). However, for backpacking trips over two days in length or for long-distance trips of lesser duration, be very selective when planning your menu, because of the limited room in your pack and the limited amount of weight you can physically carry.

While the amount of camping gear and clothing you take on any backpacking trip remains fairly constant (for example, you need a sleeping bag, the stove, warm clothes, and pots and pans for a two-day as well as a 30-day hike), the amount of food you must carry varies considerably. How much food you can fit in your pack and how much of its weight you can carry are the major factors limiting how many miles and how many days you can hike without a resupply. Generally, under ideal conditions like warm weather, you can carry no more than a two-week supply of food without a resupply. Always plan on carrying at least 1½–2 pounds of dry food per person each day in summer and at least 2–2½ pounds of dry food per person each day in cold weather. Remember that large-sized people will frequently eat twice as much as smaller people. Plan accordingly.

Backpacking food should contain the most energy and nutrients with the least amount of water in it to reduce its weight and bulk. While *dried* and *dehydrated* foods generally refer to all foods with water removed from them, they technically mean foods with a great deal (about 70–98%) of their water simply evaporated out. Examples of dried foods include raisins which are dried grapes and beef jerky which is dried beef. You can dry many foods yourself, which is a lot less expensive than buying them in a store (see page 137). *Freeze drying* is an expensive process that freezes foods at low temperatures so that their water sublimates out without changing to a liquid first. This process removes virtually all the water locked inside them, so freeze-dried foods are the lightest and least bulky of all dried foods. These drying techniques have virtually no effect on the nutrients in foods.

Cost

The cost of food is one area of your backpacking budget you can trim substantially if necessary. Specialty dried foods available from camping supply catalogs and outdoor stores are slightly lighter in weight but painfully more expensive than other kinds of dried foods readily available in grocery stores. While they're useful and often needed on fast-paced or long-distance hikes where weight and bulk are very critical, on less demanding hikes dried foods from grocery stores are ideal and much less expensive. Some backpackers obtain all their food from grocery stores, others get it all at camping stores or from catalogs, and still others complement store-bought dried foods with specialty dried foods according to their taste, weight, and cost restrictions. Generally, if you're going for a long-distance, strenuous hike, the light weight and low bulk benefits of specialty dried foods outweighs their high cost, but if you're going on an easy-to-moderate, less-than-one-week hike, you can find all the kinds of food you'll need at a local supermarket. The charts on pages 131-136 list some excellent lightweight foods readily available in almost every food store.

Food Tips

1) Eat more carbohydrates during the day for quick energy and a greater amount of fats in the evening or when not exercising, because they require a lot more time to digest. Many backpackers eat a moderate but fast breakfast to get an early hiking start, three to four carbohydrate snacks during the day to maintain their energy levels while hiking, and then a large dinner in the evening when they have time to relax and enjoy it.

2) Plan your meals so that you always have a little food left over at the end of every trip to guarantee that your meals aren't too skimpy and to provide a food supply for possible emergencies. Planning a menu so that you have neither too much food left over nor too little food during a hike is an art developed only with experience, though. Learn from your mistakes.

3) On a trip, eat the heavier, bulkier foods first and the compact, lighter dried foods last. For example, feel free to carry oranges on a ten day hike, but reduce the weight in your pack by eating them on the first day.

4) Plan dinners with a starch base (like rice, potatoes, and noodles), and add fats and proteins to it. Soups, casseroles, and stews are excellent backpacking foods because they're simple to prepare, eat, and clean up.

5) The best lunch foods are light, quickly eaten, and uncooked.

6) Test new meals and recipes at home before relying on them outdoors.

7) Carry both cooked and uncooked food for variety and flexibility. If you don't have time to cook or if a sudden storm comes up, eat snack foods not requiring cooking to ease your hunger until you can cook a regular meal under more leisurely circumstances. Examples of uncooked foods include gorp (see page 136), nuts, cheese, crackers, and peanut butter.

8) Plan a detailed menu for every hike until you become familiar with the kinds of foods available and the amounts of them you need. The larger the group of people you're hiking with, the more carefully you have to plan a menu. Always write out a menu for extended hikes and when weight is critical. On longer trips, keep a checklist of foods you have in your pack and mark them off as you eat them to eliminate surpluses at the end of the trip and shortages during it.

9) In hot weather or when sweating heavily, prevent cramps, headaches, and nausea by consuming greater than usual amounts of *salt* with your meals, but drink water whenever taking salt and never take any

if water is scarce. Carry salt or *salt tablets* in a small plastic bag, 35mm film canister, or miniature shakers available in stores. Other *spices* are light and compact so feel free to carry them at any time for seasoning your food.

10) In hot weather powdered *electrolyte replacement drinks* like Gookinaid ERG™ replace minerals as well as fluids lost when you sweat, help prevent muscle cramps, and reduce fatigue. They are much better for you than the sugar-based powdered fruit drinks which only give a brief "sugar rush" and replace lost fluids.

11) When weight is critical on short hikes it's often wise to leave your stove at home and eat prepared foods not requiring cooking. On the other hand, on hikes over four days long, it could be foolish to leave your stove at home, since the stove, fuel, and dried food could weigh a lot less than prepared dried food you can eat uncooked. For example, a five-day supply of instant oatmeal and a stove weigh less than five days' worth of dried granola, assuming you prefer the oatmeal cooked. (Uncooked, "old-fashioned" oatmeal with raisins makes a fine cold cereal!)

12) In very hot weather like the desert southwest, don't carry a stove. *Solar cook* your food by placing it in a pan in direct sunlight for a while. Although the food won't get extremely hot, in severe heat the last thing you'll want or need is a hot meal.

13) *Instant soups* provide a quick source of warmth, fluids, salts, and calories for energy. They are ideal in the late afternoon when you are tired from hiking and when the temperature is dropping.

14) An excellent method of designing a variety of trail dinners is to begin with a base of noodles or instant rice. Then add dried meat, dried vegetables, and a *seasoning packet* like taco mix or sour cream sauce. You can easily dry your own meats and vegetables (see page 137) and the seasoning packets are readily available in every grocery store.

15) Add flour, cornstarch, or instant mashed potatoes to thicken soups shortly before eating.

16) Often you'll need *fats* for cooking and to increase the calories in your food. Although *cooking oil* has a rather bland, neutral taste, it keeps indefinitely in warm weather. *Margarine,* while having a more appealing taste, spoils within a week in warm weather but lasts a long time if kept reasonably cool. Avoid *butter* since it spoils far too easily if not refrigerated. Carry your fat in a wide-mouthed, screw-topped plastic container inside a plastic bag to prevent leaks in your pack. In cold weather or when exercising heavily, increase your caloric intake by adding fat to almost any food, including cereals, sandwiches, soups, and stews.

Group Camping and Menu Planning

When camping with a group of people, it's important to choose your companions carefully, since personal problems develop easily in an out-door setting. Little problems like deciding who will sleep on which side of the tent, how fast the hiking pace should be, and who should cook dinner or wash dishes can erupt into major conflicts, especially if people have different expectations of what the trip should be like. For a success-ful group hike, proper planning is essential. Everyone in a group must be aware of and help decide the purposes of a hike and the demands it will place on the group as a whole long before they actually prepare for the hike. Another way to limit the amount of personal problems that can arise on a trip is to restrict your group size to a maximum of 8–10 people and to select a "leader" or "organizer" to coordinate the planning, preparation, and implementation of the hike. See page 185 for a discussion of methods of dealing with group organization at a specific campsite.

Planning the menu for a group of people for a weekend hike is only a minor hassle, but becomes a real chore when you have to consider weight, bulk, and cost as well as personal tastes on an extended hike. Because meal planning is very difficult for larger groups of people, it's often best to divide a large group up into sub-groups of two to four people each, especially when cooking on backpacking stoves. The following ways to arrange the food for groups work when the whole group is eating as a single unit or when it's divided up into smaller sub-groups. Choose the method that will work best with your group under your particular situation:

1) Each individual is responsible for bringing his or her own food and utensils to cook over a community campfire or several community stoves. While the responsibility for eating is placed directly on the indi-vidual, to limit complaints about the kinds of food brought and the delega-tion of cooking and cleaning chores, this method increases the group's cooking time and adds extra utensil and food weight to the group as a whole, since much of the food services are duplicated (everyone has to carry their own sugar and pans, for example).

2) If everyone doesn't mind eating the same kind of food, select one person to buy it and then divide the food's cost by the number of people eating it. This method is convenient for cooking but ignores individuals' food preferences and is a real chore for the person stuck buying the food.

3) A third method is to plan one large group dinner each evening and have everyone bring their own breakfast, lunch, and snack foods.

4) If three people are hiking for three days, have one person bring all the breakfasts, another bring all lunches, and the third bring all dinners. A variation of this is to have each person be responsible for supplying all the food needed for the whole group for one day.

5) A final possibility is to have everyone be completely self-sufficient by carrying all their own cooking gear and food. This limits the amount of group interaction and socializing at dinnertime and substantially increases the weight of duplicated gear. However, many groups of experienced hikers prefer this method since they can split from and rejoin the group at will. The added weight and loss of social interaction are small prices to pay for their independence.

Lightweight Supermarket Foods

Sources of protein:

—beef jerky
—all kinds of nuts—sunflower seeds, cashews, almonds, walnuts, coconut, etc.
—powdered milk
—cheese
—powdered eggs
—wheat germ
—canned meats—tuna, chicken spread, ham, etc.
—salami
—instant breakfast drinks and bars

Sources of carbohydrates:

—bread
—crackers
—noodles
—rice
—candy
—honey
—jelly
—instant potatoes
—dried fruits
—hot chocolate
—pudding
—jello
—powdered fruit drinks
—all kinds of breakfast cereals—oatmeal, grits, etc.

Sources of fats:

—all kinds of nuts
—chocolate bars
—margarine and cooking oil
—cheese

The charts on the following pages list ideal backpacking foods readily available in almost every grocery store.

Breakfast

Main Ingredient	Uses/Examples/Notes	Additional Ingredients
hot and cold cereal	oatmeal, grits, compact dry cold cereals try eating "hot cereals" like oatmeal cold.	wheat germ (should keep one to two weeks unrefrigerated on the trail), nuts, raisins, brown sugar, honey, jelly, margarine, powdered milk, fresh fruits and berries, dried fruits, shredded coconut, peanut butter, gorp (page 136) mince meat.
dried fruit	raw, soaked in water overnight, or cooked in boiling water until soft	cinnamon, brown sugar
instant breakfast bars	for quick starts, no cooking, no cleanup ideal trail snacks	
instant breakfast drink	drink hot or cold	powdered milk
eggs	powdered or cracked open (discard the shells, and carry in plastic containers)	cheese, powdered milk, beef jerky, canned meats, dried vegetables
pancakes	just add water mix for simplicity difficult to cook on stoves make syrup by mixing brown sugar and water in a thick paste and heating	powdered milk, jelly, margarine raisins, brown sugar, honey, cinnamon, dried fruits

Main Ingredient	Uses/Examples/Notes	Additional Ingredients
Bisquick dough	for making bisquits, dumplings, pancakes	margarine, jelly, honey, brown sugar, powdered milk, bits of cheese, cinnamon, raisins, dried fruits

Lunch

Main Ingredient	Uses/Examples/Notes	Additional Ingredients
bread	sandwiches the darker and harder the better—soft white is worst; it smashes easiest	peanut butter, jelly, honey, margarine, cheese, salami, small cans of meat or meat spreads
crackers	all kinds—saltines, corn chips, etc. sandwiches or plain store in cardboard box to keep unbroken	same as for bread
cheese	eat plain, in sandwiches, or add to casseroles, rice, potatoes, and other dishes the harder and drier, the better it keeps. Lasts one to two weeks in hot weather; longer in cold scrape off and discard any mold that forms sweats out oil in hot weather but is still safe to eat—wrap carefully in plastic cheese wrapped in wax lasts longer and is less messy in hot weather avoid processed cheese in hot weather	
dry salami	plain or in sandwiches unrefrigerated kind lasts longer	cheese, bread, crackers

Dinner

Main Ingredient	Uses/Examples/Notes	Additional Ingredients
noodles instant rice	cheese and rice macaroni and cheese noodles and rice add to soups add to bouillon cube broth buttered noodles cook in half milk, half water solution for more protein pour a hot mincemeat sauce over rice	cheese, margarine, tuna, dried meats soup, bouillon cubes, powdered milk, dried vegetables, pre-packaged seasoning mixes, mincemeat
soups	dried and powdered	add rice, noodles, or instant potato flakes for thicker soup, add bouillon cubes for more flavor, add cheese or margarine for more calories, add dried milk for creamed soup
"helper meals" "stuffing mixes"	without hamburger or stuffing alone or with dehydrated meats, noodles, rice, instant potatoes	rice, noodles, potatoes, margarine, powdered milk dried meats, dried vegetables
instant potatoes	mashed	cheese, margarine, powdered milk, bouillon cube gravy, dried meats, dried vegetables
dried meats tuna	casseroles with rice, noodles, potatoes, dried vegetables use tuna packed in oil for more calories fried in oil	bread, crackers, rice, noodles, dried vegetables
Dried vegetables	rehydrate in cold water, eat as salad	salad dressing seasoning packets

Desserts

Main Ingredient	Uses/Examples/Notes	Additional Ingredients
instant pudding	make according to directions	powdered milk
jello		
fruit cake	eat plain fry in margarine heavy in weight but high in calories	margarine
marshmallows	eat plain or toast on a fire	chocolate bars, crackers, peanut butter
popcorn		margarine, salt
dried fruit (see "breakfast" heading)		
mincemeat	cook in water until soft	

Trail Snacks

Main Ingredient	Uses/Examples/Notes	Additional Ingredients
nuts	eat plain, add to cereal, gorp, etc.	
hard candy	eat plain	
tropical (semi-sweet) chocolate	eat plain melts only in very hot weather	

Trail Snacks (*cont.*)

raisins	eat plain, add to rice, potato, noodle casseroles, cereals, sandwiches
mincemeat cubes	eat plain, add to cereals
dried fruit	eat plain, add to cereals
granola	eat by handfuls like popcorn
gorp	eat by handfuls like popcorn contains any amounts of any of these: raisins, shelled peanuts, M&M's, chopped dates, coconut, shelled sunflower seeds, dry cereal, granola, Cracker Jack. (A mixture of peanuts, raisins, and M&M's is an old-time favorite, but create your own recipe.)

Drying Supermarket Foods

The Drying Technique

The following general directions explain how to dry most kinds of supermarket foods in *a kitchen oven*:

1) Spread the food directly on cookie sheets.
2) Place the cookie sheets in your oven on low setting (about 120–170°F).
3) Rearrange the food frequently for even, thorough drying.
4) Break liquids like tomato paste into chunks or crumble into a powder when dry.
5) Store in air-tight plastic bags or containers when dried to your satisfaction.

Solar drying food will reduce your energy consumption but it takes several days of direct sunlight and periodic attention for foods to dry completely with this method. Consider initially drying food in the sun when its moisture content is highest, and then finishing the process in your oven. To solar dry food, simply place sliced foods on cookie sheets in direct sunlight. Cover with light cheesecloth if any insects are nearby. Never leave food out overnight, since it will reabsorb some of the moisture it lost during the day.

With drying, there's plenty of leeway involved between drying food too much and not drying it enough, so it's hard to make mistakes. If you think your dried food is somewhat "chunky" when rehydrated, simply dry it for slightly shorter periods of time or rehydrate it for longer time periods. You'll learn the technique rapidly with practice. In general, vegetables are dry enough when they rattle on their trays. Fruits are dry when leathery in appearance and when no moisture comes out when squeezed.

Dried foods keep at least six to 12 months when stored in your freezer. Except for meats (noted below), dried foods keep at least several weeks unrefrigerated on the trail.

Kinds of Foods

You can dry the supermarket foods listed below:

1) *Canned tomato paste and sauce*—Simply pour a can on the cookie sheets and place in your oven. When one side is dry, peel it from

the sheet, flip it over, and dry the other side. Use as a sauce in soups, casseroles, and spaghetti.

2) *Canned fruit*—Drain syrup completely before drying.

3) *Canned applesauce*—Dry like tomato paste. Makes a great trail snack.

4) *Fresh meats* (bacon, chicken, hamburger, fish, ham, beef, turkey, etc.)—Cook thoroughly by frying, boiling, or broiling, or in a crock pot. Then completely drain all fat, remove all skin and bones, cut into small pieces, dry on newspapers or paper towels (being sure to promptly remove any paper soaked with grease), stir in (optional) dry seasonings, and dry. Meat keeps unrefrigerated for at least a week in hot weather and as long as several weeks on cold weather hikes if all the fat is removed from it.

5) *Tuna* (canned in water)—Drain completely and dry.

6) *Vegetables*—Slice fresh vegetables, slightly steam or boil them to destroy their cell structure, dry, and store. The following fresh vegetables are easily dried: asparagus, carrots, green beans, dried beans (cooked before drying), bean sprouts, broccoli, cabbage, cauliflower, eggplant, mushrooms, onions, green peppers, potatoes (sprinkled with lemon juice to prevent darkening), spinach, and tomatoes. Simply dry (without boiling first) prepackaged frozen vegetables and store them. Do not dry any frozen vegetables packaged in a sauce-type mix.

7) *Fruits*—For fresh fruits, slightly steam (to destroy their cell structure), pit, slice, dry, and store. Coat noncitric fruits with lemon juice to preserve their color. The following fresh fruits are easily dried: peeled apples, apricots, bananas, berries, cantaloupes, cherries, seedless grapes, oranges, honeydew, grapefruits, lemons, peaches, pears, pineapples, prunes, plums, and watermelon. Cut and open the sections of fruits like grapefruits and oranges for better drying. Dry frozen fruits like strawberries and blueberries in their syrup like tomato paste.

8) *Mincemeat*—Dries into a pemmican food. It's great for trail snacks.

9) *Cottage cheese* (lowfat)—Dry like tomato paste.

Rehydration at Camp

To rehydrate dried foods at your campsite, simply soak them in water for 30–45 minutes. If you like to hike until dark, eat a hurried dinner, and then go to sleep immediately afterwards, rehydrate dried foods in a tight-fitting container or wide-mouthed jug as you hike the last half hour to your campsite.

Commercial and Home-Made Dryers

Several companies market *electric food dryers*. Although initially expensive, in terms of energy used they are far less expensive than the oven-drying method explained above in the long run. In addition, they're more reliable and require less attention than the oven-drying and solar methods. Quality electric dryers have removable solid (for liquids) and screened racks, a reliable guarantee, and a complete book of directions and recipes.

You can build a *homemade food dryer* by lining a large cardboard box or wooden crate that has a lid with several sheets of aluminum foil, placing an electric lightbulb inside it, and building a wooden or brick contraption to hold trays made from wire screens, cake baking racks, or cookie sheets. Poke several nickel-sized holes in the box lid to let evaporated water vapor escape (Example10-1). This method is the least expensive but least reliable of all drying methods.

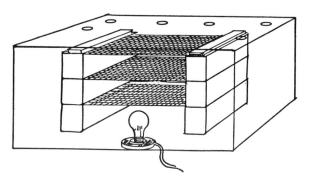

Example 10-1: Building a food dryer at home.

II

TECHNIQUES

11

Getting Started—
Planning and Preparation

Planning a Hike

Physical Conditioning

The kind of exercise you do as well as how regularly and consistently you do it determines how much it benefits you. Exercising 30–60 minutes every other day is much better than exercising for several hours each Saturday. In addition, it's far better to maintain a constant level of fitness by regularly exercising throughout the year than going on a crash exercise program just before a backpacking trip. Walking, running, bike riding, jumping rope, and cross-country skiing are excellent ways to maintain your physical condition for backpacking because they develop your *cardiovascular fitness* (heart and lungs) as well as muscle strength needed for carrying heavy loads. Never go on a long or difficult hike if you're in poor physical condition or with the belief that you'll get in shape on the hike.

Where to Hike

You can obtain information on where to hike from several sources. Regional *trail guides*, which are available from many outdoor sports stores, give ideas where to go but by no means are complete listings of possible hiking locations. *Sales clerks* in camping stores and *rangers* at parks can offer invaluable suggestions about trails suited for your skills and abilities. Local *hiking clubs* and *college recreation departments* encourage beginning trips because many people prefer learning from others more experienced than they are. *Community colleges* and *adult education programs* frequently offer backpacking classes. *Guide services* advertised in national outdoor magazines are ideal for people with various levels of outdoor experiences interested in exploring famous or remote places. As you gain experience, though, all you'll need is a *road map* to locate areas suitable for hiking. Simply find an area that looks undeveloped and interesting, buy the *topographic maps* for it (see pages 210-214 for uses), and plan your own hikes there.

With adequate preparation you can backpack anywhere and everywhere outdoors. For example, if you are unfamiliar with the Grand Canyon or with backpacking, the rangers there will direct you to a "highway trail" in the Phantom Ranch area of the park. A topographic map of the Canyon will reveal a handful of other, lesser-used trails. Yet the Canyon abounds with hundreds of miles of almost unknown, unused, unmaintained paths and thousands of miles of hikeable routes—burro trails, old Indian trails, prospecting trails, and passages obvious only from the lay of the land.

For your first few overnight trips, go to the same place to become comfortable in the outdoors and with your equipment, and plan your hikes on easy, well-traveled trails. While many experienced backpackers avoid weekends, holidays, and the crowded summer season, go at those times to increase your outdoor confidence by being around backpackers and to learn about techniques and equipment from them. Gradually, as your outdoor skills develop with experience, try other hikes in more rugged, isolated areas.

Weather Conditions

Be aware of the weather conditions for the place you will hike so you know what kind and amount of gear to pack and if you can physically deal with those conditions. *Average monthly temperature and precipitation tables* indicate typical weather conditions for a given time of year. Tables listing these values are available from:

U.S. Department of Commerce
National Climatic Center
Federal Building
Asheville, NC 28801

Request "The Monthly Averages of Temperature and Precipitation for (your state of interest) Climatic Division 1941-1970." *Five-day forecasts* indicate the specific conditions you'll probably encounter, like a sudden blizzard or a heat wave. These are available in local newspapers, from TV or radio stations, or by calling the U.S. Weather Service in the region where you'll be hiking. Above all, be prepared for the unpredictable. Cloudy conditions in a city could indicate a severe storm in the nearby mountains.

Kinds of Hikes

By definition, a *hike* is a walk with a purpose, a *day hike* is a short hike not involving camping, a *backpacking hike* is an overnight hike, *orienteering* is hiking with a map and compass, *bushwacking* and *cross-country hiking* are hiking where there is no trail, and *exploring* is hiking with no specific destination in mind. Obviously there are many possible combinations of these kinds of hikes. You can go on a bushwacking day hike close to home, you can backpack for short or long distances and camp in a different location each day, or you can set up a base camp for a few days and take day hikes from there. An interesting kind of exploratory hike is to hike and camp wherever you want to each day with no specific destination or plan in mind (Example 11-1).

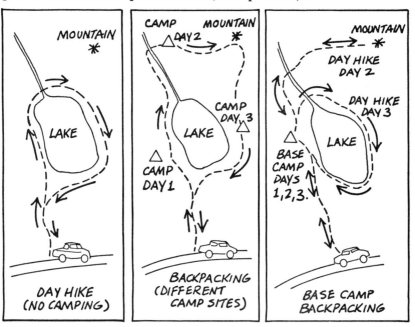

Example 11-1: Kinds of hikes.

You can plan a *circular hike* where you return to your starting point by walking in a circle, you can hike to a destination and *retrace your steps* on the same trail back to where you started, or you can do a *one-way hike* where you hike only in one direction. (See Example 11-2.) If you do that, switch car keys with a friend who is hiking the same route from the other direction and each drive the other's car home, hitchhike back to where you begin, or have a friend drop you off at the beginning and pick you up at the end of the hike.

A *car shuttle* which involves leaving a car at the endpoint and one at the beginning *trailhead* (a place where a trail crosses a road) is ideal for groups of hikers. Drive both cars to the beginning trailhead, unload the equipment and hikers, drive both cars to the endpoint where you leave one car, and return in the first car to the beginning point where you meet the rest of the group and begin hiking. Always lock your car and keep all valuables hidden. Carry the key in your pack or hide it on or near the car if you're going to switch drivers. If you drive very far to a trailhead, leave a change of clothes, comfortable footwear, water, food, and money for gas or food on the return trip in the car.

Time, Distance, and Walking Speed

How many miles you can walk in a given day depends on the factors listed below, which assume an "average" hiker with a backpack and excludes time spent resting.

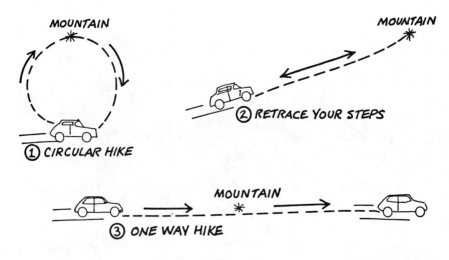

Example 11-2: Kinds of hikes.

1) *The condition of the trail.* While most people can walk about 2–3 miles an hour on a sidewalk, you'll probably hike about 2 miles an hour on a smooth, flat trail, between 1–2 miles an hour on a rough, rocky, but still flat trail, and considerably less than 1 mile an hour when traveling crosscountry through dense undergrowth.

2) *Your level of physical fitness.* A physically fit person hiking at a speed of 2 miles an hour could comfortably hike 10–20 miles in a day, while a person hiking at that speed won't go farther than a few miles if he tires quickly. Thus, when planning a backpacking trip, how long you can hike matters far more than how fast you can hike. In addition, you must measure hiking endurance in terms of the number of consecutive hiking days as well as the number of hours hiked each day. It's one thing to walk 10 miles in a day, but it's an entirely different matter to hike 10 miles a day for ten consecutive days.

3) *The weight of your pack.* Obviously, the more your pack weighs, the slower you'll hike and the faster you'll tire. Plan on walking fewer miles each day in cold weather or on longer hikes when your pack contains extra clothing and food. Plan on hiking more miles per day at the end of a trip than at the beginning when your pack contains a lot less food and thus weighs less.

4) *The amount of hiking time available.* Although you can hike anytime you want to, most people hike during the day and sleep at night. If you allow a minimum of 1 daylight hour for waking up and cooking breakfast and 1–2 daylight hours for setting up camp and cooking dinner, you'll have only about 6–7 hours of available daylight hiking time in winter and 9–11 daylight hours for hiking in summer. Deduct time spent resting and sightseeing from those figures, and reduce them further when camping in steep canyons, rugged mountains, or thick forests which block the sun's light when it's near the horizon.

5) *Elevation.* The elevation you hike at and the elevation you gain while hiking significantly affects how many miles you can walk in a day. Allow an additional hour of hiking time for every 500 feet of elevation gained if you're an out-of-shape beginner and an additional hour for every 1,000 feet of elevation gained if you're an experienced, conditioned hiker. At high altitudes, add 1–2 extra hours for each gain of 1,000 vertical feet. Also, as a general rule, remember that it takes you at least twice as long to hike up a steep trail as it does to hike down it (Example 11-3).

6) The *size of your group.* Generally, the larger the group, the slower the pace. This is especially true for groups of beginning backpackers.

	smooth trail, level ground	moderate trail, uneven ground	rugged trails, steep canyons, mountains, high altitudes
out-of-shape, inexperienced beginner	1½–2 m.p.h. 2–4 hours max. 3–8 total mi.	1–1½ m.p.h. (plus 1 hour for every gain of 500 vert. ft.) 2– 4 hiking hours 2–6 total miles	1 m.p.h. (plus 1–2 hours for every gain of 500 vert. ft.) 2–4 hiking hours 2–4 total miles
experienced, conditioned backpacker	2–3 m.p.h. 6–12 hours max. 20-30 total miles	1½–2½ m.p.h. (plus 1 hour for every gain of 1,000 vert. ft.) 6–10 hiking hours 8-20 total miles	1–1½ m.p.h. plus 1 hour for every gain of 1,000 vert. ft.) 6–10 hiking hours 6-15 total miles

Example 11-3: Approximate hiking times. (Assumes a medium-weight backpack in mild summer weather. Does not include rest stops.)

Of course, the information presented above is only a guideline to help you plan your hikes. Because individuals vary greatly in their personal hiking style, walking speeds, and physical condition, learn your own upper limits for how many miles you can comfortably walk and how much elevation you can comfortably gain in a given amount of time. Be suspect of guidebooks and trail signs that indicate distances measured in terms of hours, because they (like the guidelines listed above) are designed for "average" hikers, which in reality don't exist. Above all, recognize the difference between how far you *can* walk and how far you *want* to walk in a day.

Be flexible with your hiking plans. Allow an occasional *rest day* in case you become stormbound, get sick, or decide to stay at an attractive campsite for an extra day. Experienced hikers should allow at least one rest day for every six to ten consecutive hiking days, while average hikers should allow one for every four to six hiking days, and people with young children need at least one rest day for every two or three days spent hiking.

Hiking with Small Children

When hiking with small children, plan unpressured, easy hikes to familiar places in warm, dry weather. Consider going to an interesting lake or stream, since children are fascinated by water and can spend hours exploring there. Children enjoy camping because it's fun, so let them do exciting things, like cook, build a campfire, fish, or swim. If your children enjoy themselves the first several times outdoors, they'll jump at the chance to go camping. If their first experience of camping is huddling inside a leaking tent in a storm, it'll undoubtedly be their last camping experience. You should only take young children hiking if you're an experienced hiker and if your children truly want to go. Only take children on overnight camping trips after they've day-hiked and camped in their backyard several times. Experience is the best teacher.

Selecting and Packing Equipment

The Checklist Method

The checklist method of packing your equipment is simple and foolproof. Simply read down the checklist on page 150 one item at a time and collect the things you'll need for a hike in a pile in a room. Only collect the things you'll need for that specific hike and ignore everything else on the list. For example, you won't need a swimming suit for a winter hike, so ignore it; however, you'll need more than one jacket or coat, so be sure to pack enough of them. One list is all you need to use for every hike. Simply pack what you need and ignore what you don't need. When everything you've collected for your hike is piled on the floor in front of you, try to reduce the weight you'll have to carry by eliminating more things you don't really need. Ask yourself if you really need two jackets instead of one, four pairs of socks instead of three, and a tent if tarp will suffice. "When in doubt, leave it out," experienced backpackers say.

As a general rule, it's best not to carry a pack weighing more than one-quarter to one-fifth of your own weight. (Experienced long-distance hikers, however, often ignore that rule. 118-pound Eric Ryback, for example, carried a pack frequently exceeding 80 pounds on his 2,500-mile Pacific Crest Trail hike.) If you're a beginning backpacker, start with light loads on easy hikes in mild weather and gradually build up to longer trips and heavier loads. A pack for a two- or three-day hike in mild weather shouldn't weigh more than 20–30 pounds, while one loaded for a three- to five-day hike under similar conditions should weigh about 25–35 pounds, and one for a weeklong trip will probably weigh about 35–40 pounds. Those guidelines vary considerably depending on the weather conditions and how many extras like camera equipment you carry. You can weigh a loaded backpack by simply setting it on a bathroom scale, or by wearing it, standing on the scale, and then subtracting your weight from the measured total.

Equipment Checklist

Shelter

pack
sleeping bag (stuff sack)
ground cloth
tent (fly, stakes, poles)
tarp (visclamps, rope)
bivouac bag
hammock
foam pad/air mattress
vapor barrier
radiant heat barrier

Clothing

pants (long and short)
shirts (long and short)
t-shirts
underwear
long underwear
heavy coats/sweaters
swim suit
sun hat
windbreaker
parka
gloves/mittens/liners
wool hat
vest
vapor barrier clothes

Rain Gear

poncho/parka/etc.
rain hat
pack cover
chaps/rain pants

Feet

boots
sneakers
winter boots
camp moccasins/sneakers/sandals
booties
gaiters
socks

Health

towel/bandana
toothbrush/paste/dental floss
comb
razor
mirror
soap
toilet paper
lip balm
insect repellent
sun tan lotion
foot powder
water purification tablets
first aid kit
  rubbing alcohol
  sunglasses

Cooking

canteen
pots and pans
silverware
cup
stove

stove supplies—funnel, eyedropper, gas cap pourer, nozzle cleaner
fuel
matches
firestarter
scrub pad
food
pocket knife

Other

identification
compass
maps and guides
fire/camping permits

flashlight
extra batteries/bulb
extra eyeglasses/carrying case
repair kit
survival kit
money
pen and notebook
watch
alarm clock
rubber bands/plastic bags
rope
rope for belt
recreation—books, games, fishing tackle, camera equipment, etc.
candle, candle lantern
trowel

Repackaging Food

There are several ways to repackage your food to reduce its weight and bulk. You can pack each food item separately to save packing time at home and allow more versatility on the trail. For example, by packing your oatmeal, raisins, and sugar in separate bags, you can eat cereal with no raisins or with plenty of raisins, or you can eat the raisins plain. A second method of repackaging food is to pre-mix everything for the same meal together to avoid mixing foods on your hike and searching through your pack for three or four bags of individual ingredients. For example, make packets containing oatmeal, raisins, and sugar mixed in correct proportions for each meal. That way you simply find one packet in your pack, cook it, and eat it (Example 11-4). Similarly, when packaging

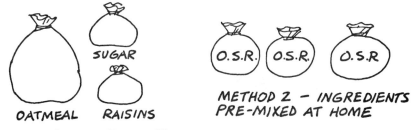

Example 11-4: Ways to repackage food.

ingredients like powdered milk, bag the powdered milk in one large bag or make small individually wrapped servings of it which are then stored in a large bag that contains all the individual bags.

With the exception of fragile foods like crackers, remove as much as possible of the original cardboard, glass, and can packaging that came with the food when you bought it to save weight. Put loose powders and dry foods into plastic containers (Example 11-4a) or plastic bags and put an extra bag around any that could break open. Squeeze as much air out of them as possible before sealing them with a rubber band. In cold weather

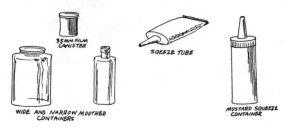

Example 11-4a: Kinds of containers useful for storing foods.

when rubber bands are a hassle, simply twist food bags closed and pack them carefully in your pack. Don't use wire twist ties because they'll puncture holes in other bags in your pack. Put all gooey foods like honey and peanut butter in plastic squeeze tubes or containers with screw-on lids to prevent leaks and with wide openings for easier cleaning, but avoid squeeze tubes in cold weather when their contents harden and they freeze shut. The openings on wide-mouthed plastic containers should be large enough to easily let your spoon or knife exit it when loaded with a gooey food like peanut butter. Narrow-mouthed containers and empty mustard squeeze bottles are ideal for storing powdered foods like sugar and dried milk. You could have a real disaster if a plastic bag of those foods broke inside your pack. Carry small items like bouillon cubes, salt, and water purification tablets in 35mm film canisters or drugstore pill bottles. Always label everything you can't just look at and identify and pack recipes or cooking directions with food when necessary.

On long hikes or when you have a limited amount of room in your pack, consider *bagging* your repackaged food to better organize things in your pack, to reduce the extra weight of the containers, and to eliminate the hollow bulk of containers when empty. There are several ways to do this (Example 11-5). One way is to group all the breakfasts, lunches, and dinners in separate plastic bags or different-colored stuff sacks. Nylon mesh stuff sacks are ideal because you can easily see what's inside them.

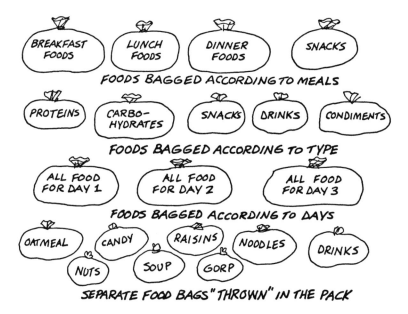

Example 11-5: Ways to bag your food.

Another way is to separate your food into large bags of proteins, carbohydrates, snacks, drinks, and condiments. A third way is to group the food according to which day you'll use it and bag it accordingly. If you don't like to group your food in any of the ways mentioned above, you can simply stuff the individual food bags and containers in your pack wherever they fit and worry about finding what you need when you need it.

A fully loaded pack provides a remarkable amount of insulation. When carrying perishable foods like eggs, fresh meat, and fresh vegetables, store them in plastic bags inside a bundle of clothes in the middle of your pack. Even ice cream will stay frozen for several hours in summer when packed this way.

Packing a Backpack

1) Pack soft items against your back when carrying day packs or internal frame packs so sharp objects won't dig into your back.

2) With a framed backpack, put the heaviest items high and inside against your back and the lightest things low and to the outside of the pack away from your back. When carrying an internal frame pack or while showshoeing, climbing, cross-country skiing, or scrambling up steep slopes, pack the heaviest things inside against your back for better balance (Example 11-6).

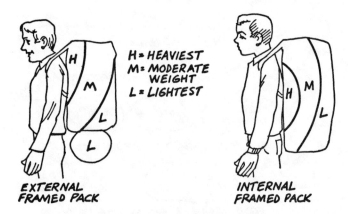

Example 11-6: Packing a backpack.

3) Pack the things you'll use frequently or need quickly, like a raincoat in stormy weather, a canteen on a hot day, and an extra sweater in cold weather, in easily accessible outside pockets or at the top of your pack, and put little-used things like the ground sheet or extra food at the bottom of your pack.

4) Always put smaller, easily misplaced items like the repair kit in the same pocket of the pack. After awhile, you'll instinctively know where they belong and will find them quickly when needed.

5) Balance the pack as you load it. If you put a heavy canteen on one side, put something equally heavy on the other side so it won't sway as you walk.

6) If you have to carry a large pot or bucket for cooking, fill it with clothing or food and put it inside your pack. Never tie it to the outside of the pack because it'll swing back and forth causing a loud noise, un-balancing the pack, and snagging on brush.

7) Carry your sleeping bag inside a plastic bag in a stuff sack to keep it dry in wet weather.

8) Keep your clothes dry in plastic bags inside your pack, especially if you have no waterproof pack cover over the entire pack.

9) Use clear plastic bags for waterproofing or storing things in, so you can see what's inside them.

10) Some people with uncompartmentalized pack bags put their clothes in one colored stuff sack, the food in another, and their camp

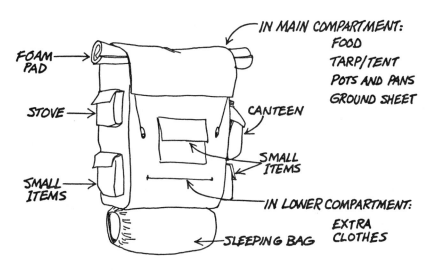

Example 11-7: How to pack a backpack.

equipment in a third, different-colored sack for better pack organization, but that takes up more room in the pack and could cause it to sway back and forth as you walk, since you can't stuff odds and ends in tiny corners to save space and balance the pack better.

11) Pack bulky clothes like sweaters and shirts by rolling them up and securing them with rubber bands.

Sharing Group Items

Everyone in a group of hikers should carry their fair share of community gear like the food, shelters, and cooking supplies. Stronger and more experienced hikers should carry a little more, while less experienced hikers should carry slightly less. Divide the community gear up according to its weight and bulk. If someone has a very small pack, have him carry a few of the heavier items, while people with larger packs should carry a lot of the lighter, bulkier things. Distribute the food so that everyone's pack lightens evenly as it's eaten. For example, in a group of three people, it's better to have one person carry all the breakfasts than to carry all the food for the first day of a three-day hike, because, after that first day is over, the other two hikers still would have fully loaded packs while his would be considerably lighter.

At the End of a Trip

Place all your equipment from the trip in three piles—what you used frequently or needed for safety, what you used occasionally, and what you didn't use at all. While some things like the first aid kit are necessary even if you don't use them, consider not carrying anything in the unused pile on your next trip. For example, ask yourself if you really needed that extra t-shirt or pair of socks. If you do this self-evaluation consistently at the end of every trip, you'll gradually make fewer bad judgements when planning and packing equipment.

Carefully store camping gear so it'll be ready the next time you need it. Air out and dry out your gear so it won't rot or mildew. Store all down garments unrolled if possible. Repair anything that tore or broke. Don't keep any food in your backpack, because it'll absorb food odors that could attract animals on the trail or at home. Store all your camping gear odds and ends like your pocket knife and compass in the pack or in a box so you'll know where they are when you need them again.

12

Hiking Techniques

The best way to avoid trouble on a hike is to be prepared for it. Always tell a reliable person where you are going, when you expect to return, and the names, addresses, and phone numbers of everyone in your group. When you tell someone you plan to return by a certain date and time, be sure you return by then so they don't become upset and begin looking for you. Never change your hiking plans after telling someone where you are going because if you need help, rescuers will look for you where you said you would be and not where you are. Before hiking in a park, get all necessary camping and fire permits and file a hiking plan with the rangers.

How to Walk

Immediately before beginning a hike, remove a few layers of clothing, since your body will quickly warm up when walking; it's better to begin a hike feeling chilled than to be comfortably warm at the start of a hike and overheated and sweaty 10 minutes later. Then do several toe touches, deep knee bends, jumping jacks, or other calisthenics to let your heart, lungs, and muscles adjust to the strain you'll soon place on them. Begin hiking slowly at first, and within minutes you'll fall into a smooth, comfortable walking rhythm. You'll walk slower and more deliberately when carrying a backpack but don't think about how you're supposed to walk as you hike. Walking with a pack comes as naturally as walking without one.

Maintain a steady hiking rhythm by keeping your heartbeats and breathing at constant, moderate levels. Don't walk so fast that you gasp for breath and your heart feels like it's going to jump out of your chest. If you're breathing so hard that you can't talk comfortably, you're hiking too fast. Also, it's better to maintain a slower, steady rhythm throughout the day than to travel in fast spurts like a jackrabbit and rest around each bend in the trail. Try to maintain your energy output at a constant, comfortable level.

Whenever possible, walk over or around objects like logs and rocks instead of stepping on and then over them. You burn up as much energy stepping up 1 vertical foot as you do walking 10 feet on flat ground (Example 12-1)

Example 12-1: In terms of energy used, 1 vertical foot equals 10 horizontal steps.

Hiking Uphill

Maintain a steady pulse and breathing rhythm when hiking uphill, even if you're just putting one foot an inch in front of the other. On steep uphill climbs, walk with a *rest step* by locking your knees and momentarily pausing with each step. Use your arms to push off your legs for more lift up particularly steep climbs (Example 12-2).

When you feel like you can't go any farther, force yourself to keep walking by saying "I'll rest when I get to that tree," or "I'll eat a candy bar when I get to that flat rock." Then reward yourself with food, water, or a rest after you've hiked the specified distance. When you've tried the rest step and the rewards and you're truly tired, take a rest. Don't fight with the trail and with yourself all the way to the top of a mountain.

In canyons and mountains, you can judge your position and the distance you still have to hike to reach the top by looking directly across

Example 12-2: Using your arms to help climb over steep obstacles.

the canyon or at another nearby mountain (Example 12-3) or by calculating the vertical distance involved with a topographic map (see page 211).

Hiking Downhill

While downhill walking puts much less strain on your lungs and heart, it shocks your leg muscles, is hard on your knees, causes blisters, and is more dangerous than uphill walking because it's easier to stumble and get hurt. Always be very alert when hiking downhill. Before walking

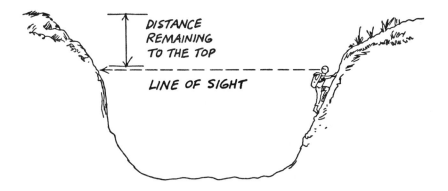

DISTANCE REMAINING TO THE TOP

LINE OF SIGHT

Example 12-3: Estimating the vertical distance you've hiked.

on a long downhill section of trail, tighten your boot laces or add an extra pair of socks to prevent foot slippage inside your boots which causes blisters.

Although you can slowly run down smooth and not very steep trails as long as you can safely stop anytime you have to, resist the temptaion to charge downhill out of control. You won't jar your feet or knees so much if you run in a flowing motion carefully using your legs to propel your momentum down the trail instead of as shock absorbers slowing you down with each footstep. Run stiff-legged downhill in loose snow and sand so you'll slide with each step, but be careful you don't snag a foot on a log or rock buried under the surface. Never run downhill on loose gravel and rocky trails where there's a great chance of tripping or sliding out of control, and never run downhill if your leg muscles are tired, if you feel it's too dangerous, or if you're in a hurry.

Rest Stops

Some people like to rest 5 minutes each hour at regular hourly intervals throughout the day. These are the people you'll see walking past a beautiful overlook because it's not time for them to rest yet. Other backpackers rest whenever they feel tired, which is often many many more times than they physically need to. They're the ones you'll see sprawled out along the trail trying to absorb as much rest as they can before their twentieth attempted "final assault" up a mountain peak. Fortunately, for the majority of backpackers that lie somewhere in between those extremes, there are some practical rules that apply to rest stops.

If you're hiking at a comfortable, moderate pace and are in good physical condition, you'll need very few rest stops during the course of a hiking day, but feel free to rest whenever you're tired or want to enjoy the scenery. Don't rest for more than 5–10 minutes at a time if you're doing any serious hiking, though, because your muscles will tighten up and it'll be much harder to begin hiking again. If you're very tired, rest often but for no more than periods of 2–3 minutes, which is long enough for your heart and lungs to catch up with the demand placed on them and to cleanse about 50% of the lactic acid (the waste product of muscular activity and the thing that makes your muscles feel tired) out of your muscles. Resting for longer than 3 minutes only reduces the lactic acid in your muscles by a few more percentage points (only a very long rest like overnight sleep will clean out almost all of it), so longer rests won't restore significantly more strength to your tired muscles.

Example 12-4 shows the ways to rest described below:

① REST YOUR PACK
ON A ROCK

② LEAN FORWARD, LOCK YOUR KNEES,
AND SUPPORT YOUR UPPER BODY
WITH YOUR ARMS

③ LAY ON THE GROUND WITH YOUR FEET
ELEVATED ABOVE THE REST OF YOUR BODY

Example 12-4: Ways to rest.

1) Find a flat rock or a log about 2–3 feet above the ground and sit on it with your pack resting on it but still strapped to your back.

2) To catch your breath when hiking up a steep hill, simply bend over at your waist, lock your knees, keep your arms straight, and rest them on your knees. A variation of this is to prop your arms against a cliff, tree, or large boulder.

3) Remove your pack, lay on the ground, and prop your feet up against a rock or a tree. This is an excellent way to refresh your legs and reduce their natural swelling from walking around all day.

4) Remove your pack and walk around or do simple stretching exercises to loosen up stiff muscles. It's often best, however, to rest without removing your backpack, since it's a real chore to put on again when you're tired.

High Altitude Hiking

Most people notice a difference in their pulse and breathing rates when the altitude change from their home to their hiking area is greater then 3–4,000 feet. Thus, let's define high altitude as any elevation that's greater than 3–4,000 feet above what you're used to. For example, by this definition, 4,000 feet is a high altitude for someone living at sea level, and 10,000 feet is a high altitude for someone living at 6,000 feet. If you don't hike frequently at high altitudes or if you're in poor physical condition, hike very slowly for the first three or four days of a long hike so your body can adjust to the difference in elevation. Don't plan strenuous, short high altitude hikes unless you're in excellent physical condition. If you feel a headache, unusual fatigue, or nausea when hiking at a high altitude, slow down and relax for a while, because your body's telling you that you're working too hard. If those symptoms persist, you could be suffering from some kind of altitude sickness (see page 251).

Since your body does most of its emergency adjustments to high altitudes in the first two to four days there, try to delay intense physical activity as long as possible after arriving at a high altitude. For example, leave home in the evening on the first night out and sleep at your high altitude trailhead, pack your food there, or day hike for a day or so before beginning a rugged backpack trip to give your body time to adjust to the elevation change.

Hiking at Night

Hiking in the darkness has its own special dangers, but its rewards are unforgettable. Sooner or later you'll want to hike up a mountain in the dark for a view of the sunrise from its summit, hike down from a mountaintop after the sun has set, or walk along a trail in the moonlight. Suggestions for walking at night follow:

1) When using a flashlight, shine it in short bursts of 1 second every 5 seconds or so to preserve its batteries and your night vision. You'll easily remember the trail conditions in front of you when the light's off.

2) Batteries last longer in cold weather if kept warm in your pockets when not in use. Lithium batteries last longer and are far more reliable than standard carbon-zinc or alkaline batteries.

3) Nurse more light out of "worn out" batteries by letting them rest unused for a while.

4) Carry spare bulbs and batteries if you plan to hike at night.

5) When bushwacking, sweep your hands around in front of you to

check for tree branches or obstructions in your way. Be especially careful that a stray branch doesn't poke you in an eye.

6) Learn to feel the trail with your feet. With a little practice, you'll immediately know when you wandered off it. Trails are almost always more packed down, have fewer loose rocks, leaves, and sticks, and are slightly lower than the surrounding ground surface.

7) To conserve batteries, individuals in a group could walk single file with every second or third person using his flashlight. The people with the lights should shine them in front of the person in front of them so more than one person can see with the light from one flashlight (Example 12-5).

Example 12-5: Five people hiking with two flashlights.

8) Preserve your night vision. Never look directly into your flashlight beam or shine it in another person's eyes. Try covering a bright light with your fingers to restrict its beam.

9) Consider using a *headlight* that attaches to your head and a battery pack that attaches to your belt if you need both hands free. Avoid carbide miners' lamps which burn out in wind and rain and are a fire hazard in dry conditions.

10) Try hiking without a flashlight if the moon is up, if the stars are bright, or if your path is clear. Once you begin using a flashlight, your natural night vision is ruined and you are committed to using artificial light to see. If you plan to hike with no lights, sit in the darkness or walk very slowly for 15 minutes or so to fully develop your night vision.

Following a Marked Trail

Most well-used trails in popular areas are marked for identification in several ways (Example 12-6). *Blazes* are painted marks or plastic disks on trees. In popular areas, different-colored blazes signify different trails.

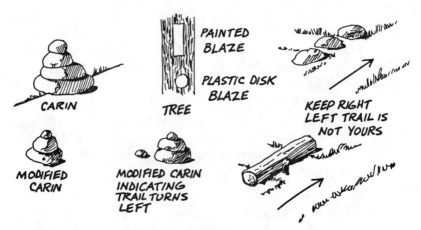

Example 12-6: Kinds of trail markings.

Cairns are large piles of rocks marking trails in treeless areas like deserts or above timberline. *Modified cairns*, which are small piles of three or four rocks, also indicate a trail. A single rock next to a modified cairn is like an arrow pointing in the trail's direction. If the trail you're following is constructed properly (and many aren't), from any given point on it you should be able to see at least one of the just described markings. Logs laid or rocks piled across a trail at a junction indicate that that is not the correct way to go.

If you haven't seen any trail marking on a marked trail for awhile, there are several ways to tell if you're still actually on a man-made trail and not on an animal trail. (Of course it could be the wrong trail. See Lost, page 272). It's undoubtedly a man-made trail if nearby trees or branches show saw or axe marks. Rocks polished smooth by frequent foot traffic indicate a well-used path. Compare rocks both on and off the trail. A path that's deeper than the surrounding ground layer signifies that the trail was worn into the soil by frequent use, while a terrace on a hillside was probably constructed there at one time (Example 12-7). In addition,

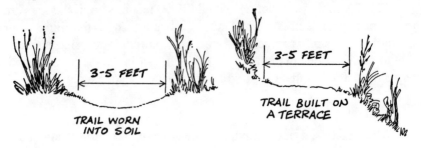

Example 12-7: Ways to identify a hiking trail.

game trails are usually very narrow, hard to follow, and littered with animal droppings, but since they frequently crisscross hiking trails, it's not hard to begin wandering on one by mistake.

Off-Trail Hiking

Bushwacking off developed trails offers its own special rewards and its unique pains. Generally, you'll enjoy off-trail walking if you're an experienced backpacker and are prepared for its challenges. When hiking cross-country, wear an internal-framed pack, because it snags on brush a lot less than an external-framed pack does, wear very sturdy hiking boots, and be careful of snakes in thick undergrowth. Allow about ½–2 hours for each mile hiked, depending on how rugged and overgrown the terrain is. When hiking cross-country, you often won't know your exact position but you must always know your general location, so be skilled in map reading, compass use, and survival skills before setting out on any off-trail hike. Specific cross-country hiking suggestions, many of which apply to trail walking as well, are listed below:

1) Don't gain or lose altitude unnecessarily.

2) Walking on ridges is often easier than walking through valleys which have obstacles that include marshes, thick forests, and lakes. Eventually you'll have to hike to lower elevations for water though, since, except for lingering snowpacks, water is scarce on exposed ridges.

3) When gaining or losing elevation, gradual slopes are easier to hike on than steep ones.

4) Carefully check your map before walking anywhere, because the shortest route is not necessarily the best one. Obstacles like swamps and cliffs could separate you from your destination.

5) Hike in snow early in the day before it softens up from the sun.

6) Hike on a *scree slope*, which is a steep slope covered with pebbles, exactly as if it was snow. Be extremely careful when hiking on a *talus slope* consisting of large boulders, though, especially if the rocks are wet.

7) Use game trails and old logging roads whenever possible, because they follow the easiest routes around obstacles and to water sources (Example 12-8).

8) When hiking in a dry wash or canyon, it's less tiring to walk on slickrock or boulders than in deep sand. Also, try short-cutting the bends of a meandering wash to reduce the distance you're hiking.

9) Walking backwards pack-first through sections of thick brush is much easier than going through face-first.

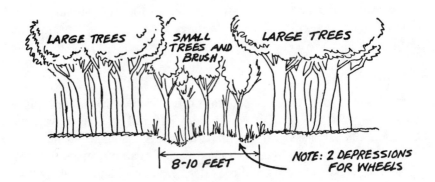

Example 12-8: An overgrown logging road.

Walking in a Group

It's a good idea to have one person designated as a lead person and another as a tail person in a group of inexperienced backpackers. The person who knows the route goes first, while the most experienced backpacker who should carry the group first aid kit, hikes last. For safety, everyone else should stay between those two people and should be able to see the person in front and the person behind them at all times. However, it's best not to hike closer than 10–20 feet from another person to avoid flying branches, startled snakes, and sudden stops, and to see more of the scenery than the hiking boots on the person in front of you. Everyone should know the color of the trail blaze your group is following (if appropriate), and the group should pause at every trail junction so no one takes the wrong turn. When hiking in a group, it's a good idea to pause for a moment in the first 30 minutes following breaking camp so everyone can re-adjust their clothing and tighten boot laces.

While groups of beginning hikers should stay close together for safety, individuals in a group of experienced hikers should walk at their own pace whenever possible. Putting slower hikers in the front of a group to slow the group down wears out the faster hikers because they're forced to walk slower than their normal walking speed, while forcing slower hikers to match the speed of faster people tires them out. If you're an experienced hiker, it's better to walk alone and meet your group at a specific destination than to force yourself to match someone else's pace. Remember though, that at times everyone needs to stick together for safety. When in doubt, stay together—but whenever you can, hike at your own speed.

Hiking Alone

Compromises are to group hikers what independence is to solo hikers. You don't have to compromise anything with anyone when you hike alone. You can walk at your own speed, sleep when you want to, eat whenever you're hungry, and do only the things that you yourself truly want to do. While solo hiking involves risks somewhat greater than group hiking, the key to hiking alone in confidence and safety is to recognize those risks and know how to deal with them. Know first aid for snakebites before hiking in the desert. Know what to do if you meet a bear on a wilderness trail. Take care of your body so you don't have a heart attack miles from help. Don't take unnecessary chances by climbing mountains, wading swift rivers, or camping in winter if you can't deal with any problems that arise yourself.

When you first hike alone, you'll confront the barrier of fear that separates your familiar world with the world of the unknown. Every strange sound will scare you and every shadow will become an unseen danger lurking along the trail. After a day or two, you'll begin to accept the fear of being alone, and then gradually, by about the third or fourth day alone, you'll come to understand that that barrier of fear was only imagined in your mind.

Eventually, however, and depending on how long you're hiking and whether you're coming to the outdoors because of what it offers or are running from something you left behind or can't find in your regular life, you'll confront a second but very real and much more immense barrier. It's one thing to go on a short solo hike as a break from your job, for a change of pace, or simply to get away from it all for awhile, but it's an entirely different matter to backpack alone for more than just a few days. After overcoming the barrier of fear of the outdoors, you'll begin to see yourself locked in a prison of solitude. You'll have the urge to talk but no one will be there to listen, you'll want to share but will wonder how you can share alone, and you'll want to be with someone for company but can't forget that you're alone. A feeling of loneliness—terrible, painful loneliness—will surround and engulf you. Indeed, then you'll realize that Abbey was right when he said "Somewhere in the depths of solitude beyond wildness and freedom, lay the trap of madness."* You'll only fully understand the deep spiritual and personal significance of solo hiking, though, when you confront that feeling of complete and total solitude.

---

*Edward Abbey, *The Monkey Wrench Gang*, page 106.

Hiking Obstacles

Climbing Up or Down Steep Slopes

You may need to climb up or down steep slopes, especially when traveling cross-country in very rugged terrain. Unless you're properly trained in technical rock climbing techniques, use ropes only to haul your pack up or down a steep drop-off and not as safety lines for supporting people. Some simple climbing suggestions are listed below:

1) Before beginning, remove clothing that restricts your movements, secure loose pack straps or shoelaces so you won't trip on them, and empty your pockets so nothing pokes you or constricts your legs.

2) When climbing, always support your body on three points of contact with the ground by moving only one hand or one foot at a time. Test all hand- and footholds with as much weight as possible before using them to support your full weight.

3) Keep your body weight centered directly over your feet. Don't lean into the slope. (Example 12-8a).

4) Use your hands as anchors to secure your position and let your legs support your body's weight. Lower your arms occasionally for better blood circulation. Never put your hands into hidden crevices in snake or scorpion terrain.

5) Keep your eyes scanning ahead for possible routes and holds. Don't look down if scared of heights.

6) If you dislodge a rock, yell "Rock!!!" to warn anyone below you. Never climb immediately behind or below another person. Either climb off to your partner's side or wait until he has reached the top or bottom before beginning.

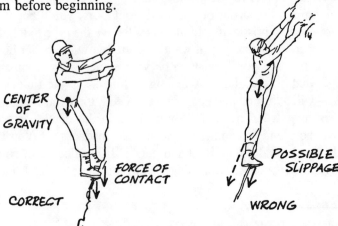

Example 12-8a: Proper body position when climbing.

7) Keep your hands and boots dry. Be extremely careful when climbing on wet rocks.

8) Try to rest your muscles and regain any lost composure or confidence at secure spots on your climb. Avoid making a "last dash" up or down the rock face to get to your destination.

9) Climbing on loose, crumbly rock is extremely dangerous. Always avoid it.

10) Usually climbing down is harder than climbing up, since you can't see hand- and footholds as easily. Always be sure you can climb down what you climbed up and up what you climbed down.

11) Don't climb with gloves on your hands unless the weather or the rock face is very cold. Finger contact with the rock gives you greater sensitivity and friction for more support.

12) Relax. Let your body flow with the rhythm of the rock and the climb. Don't fight it.

13) Don't be a hero. If the mountain is too steep, give up.

Basic Climbing Techniques

The skills described below are elementary technical rock climbing techniques which you may need to use to get around obstacles when off-trail hiking. All can be used by hikers without formal rock climbing training and without any safety gear like climbing ropes and carabiners. None should be used without common sense and a deep awareness of what your own personal climbing limits are. Personal climbing limits include fear of heights, muscular strength and endurance, level of fatigue, prevailing weather conditions, and the like. I recommend that you practice these techniques very close to the ground before relying on them on a hike. Above all, remember that this section teaches you climbing skills you can use to get around short, steep obstacles. It is not designed to teach you the art of technical rock climbing.

Climbs are classified in three general categories—friction climbs, holds, and opposition climbs. They are described below.

A *friction climb* (Example 12-8b) is when the friction between your boot and the rock supports your weight. You should use this when walking up slabs of rock that are tilted but not vertical. The *critical angle*, the angle at which friction is no longer effective, depends on the quality of soles on your hiking boots, the amount of moisture, dirt, and plant growth on the rock surface, and how rough the surface is. You can friction-walk up very steep pitches if the rock is rough, dry, devoid of lichens, moss, pine needles, and other debris, and if you have quality soles on your boots. When friction climbing, you should try to have a large area of

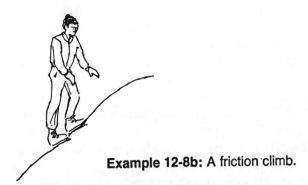

Example 12-8b: A friction climb.

contact between your boots and the rock, and should keep your center of gravity out over the rock and not in against it so your feet won't slip out from under you.

To climb up many cliffs you need *hand-* and *fingerholds* to grab onto and *toe-* and *footholds* to stand on (Example 12-8c). An *edge climb* (Example 12-8c) is when you use the side of your foot in a foothold. Whenever possible, use the big-toe side of your boots in the hold for the most support with the least muscular effort.

Opposition techniques involve using the engineering concept of opposing forces. There are four categories: jams, chimneys, laybacks, and underholds.

A *jam* (Example 12-8d) is when you stick a part of your body in a crack and then twist it so that it sticks in place. Examples are finger jams, hand jams, leg jams, and foot jams, all of which depend on the size and shape of the crack.

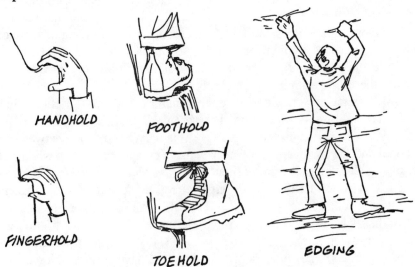

Example 12-8c: Kinds of holds.

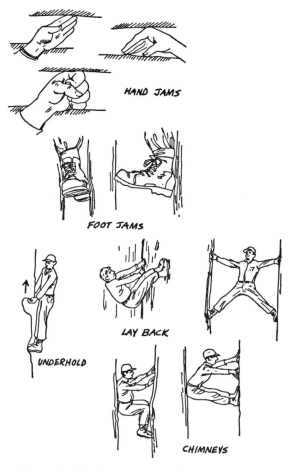

Example 12-8d: Kinds of opposition holds.

A *lay back* (Example 12-8d) is useful on vertical cracks that have one edge sticking out farther than the other. With this, you place two hands on the edge of one side of the crack and your feet on the opposing, protruding side. Your arms should be locked to relieve the strain on your muscles. Then, while hanging back and down, slowly move your hands and feet up the crack one at a time.

With an *underhold* (Example 12-8d) you apply strong upward force with your hands in a crack or indentation to support your body.

A *chimney* is a technique used to climb a rock crack larger than your body. The general idea with this maneuver is to support your weight on your hands, feet, back, or butt which are braced against each side of the crack. By altering the pressure on those parts of your body, you can inch your way up or down the crack.

Crossing a Stream

When crossing a *shallow stream*:

1) Cross it barefoot if the bottom is sandy and the water clear to keep your boots and socks dry for hiking on the other side.

2) If there are sharp rocks on the stream bottom or if the water is cold, remove your socks to keep them dry and wear your boots across it to protect and insulate your feet.

3) Build steppingstones with rocks or logs to cross narrow streams.

4) If you know you'll be crossing streams frequently on a hike, carry lightweight sneakers for that purpose.

5) If the water is more than knee deep, cross it in short pants or no pants at all. If you cross in rolled-up long pants, you could lose your balance trying to hold up long pants that are unrolling in the middle of the stream.

6) Walk with a shuffling and not a plodding step for the most stability.

7) Cross faster water facing upstream and use a walking stick for more support.

8) Always unhook your waist belt when crossing any stream so your pack can't pin you underwater if you slip and fall in.

When crossing *deep or very fast moving water*:

1) Cross mountain streams in the morning when snowmelt runoff is the lowest.

2) Repack your gear into plastic bags for protection in case your pack gets wet or you have to float it across the stream.

3) In fast and deep water, it's best to cross at an angle facing upstream or downstream. The faster the water, the more you need to angle up or down the stream. Walking straight across is the most dangerous, since the current can easily sweep you off your feet.

4) Cross in groups of three with the heaviest person in the middle for support in very rough water. Lock arms or hold on to each other or to a log for additional support (Example 12-9).

5) Use a rope belay for additional support, but never tie yourself to the rope because if you slip, your pack or the water pressure could trap you underwater. With a rope belay, travel downstream and across the stream. You'll need an extra long piece of rope for a solo belay crossing, since it must be doubled to be retrieved (Example 12-10).

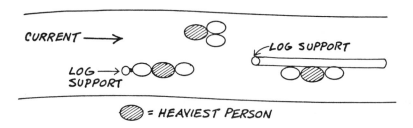

Example 12-9: Crossing a fast or deep stream.

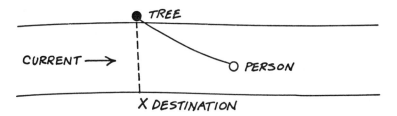

Example 12-10: Crossing a fast or deep stream with a rope belay.

Be careful of thin ice when crossing *frozen streams*. Generally, ice is thinner in the middle of a stream and on the outside of its bends, but springs and water currents could cause weak spots anywhere in the ice. Always carry a long, strong pole for support in case you fall in.

Drinking Water

When water is scarce, drink a little at a time but drink as often as needed. Try chewing gum or sucking on a pebble to help keep your mouth moist. Always drink some water when you eat, since your body needs water to digest food—and never eat anything except juicy fruits when you have no water to drink. Because you'll need over a gallon of water a day for drinking and more for cooking, washing dishes, and brushing your teeth in dry climates, carry all the water you'll need in those areas and use it sparingly until you find more along your hike. Don't rely entirely on maps to locate sources of water because if they're old, water tables could have dropped and waterways could have dried up. Rely on people like park rangers and friends or up-to-date guidebooks you can trust. Since a gallon of water weighs 8 pounds, you'll have a heavy pack when you carry all your own water on a hike in dry areas.

When plenty of water is available, drink as often as you want but not as much as you desire. Small sips are better than big gulps so you don't get a waterlogged stomach. Always carry at least a pint of purified water (see Purifying Water below) in your canteen because safe drinking water is seldom readily available all the time.

Because it's easy to dehydrate *in winter*, always drink plenty of water in cold weather, especially when very active or while eating. Never eat large amounts of snow or ice, since your body uses a tremendous amount of energy converting them into useable water and since swallowing very cold water causes stomach cramps in many people. Always melt the snow first (see below) and drink from a warmed canteen if possible. On a cold day, drinking cold water is not as chilling as eating ice, but it's not nearly as good for you as drinking a hot drink. In summer or on a hot winter's day, though, drinking ice-cold water or eating snow is as harmless and as refreshing as eating an ice cream cone.

Purifying Water

Only a few short years ago, backpackers considered any water they found away from human or domestic animal habitation safe to drink. Unfortunately, because the size of undeveloped areas is shrinking rapidly and many more people are camping in the outdoors with unclean practices, almost all outdoor water is too polluted to drink. The spreading of a microscopic protozoan called *Giardia* is an especially recent occurrence affecting traditionally safe wilderness camping water supplies. Symptoms of *giardiasis* (and other polluted water-caused sicknesses as well) include diarrhea, nausea, upset stomach, and headache. While these symptoms usually appear shortly after drinking polluted water, they could appear up to three weeks after your hike. For your own safety, purify all water you use unless you are absolutely sure it's safe. Even purify water used for brushing your teeth.

There are four major ways to purify water:

1) You can *boil* water for at least 5–8 minutes (longer at high altitudes), which is relatively convenient if done while cooking a meal or if you're not in a hurry, but troublesome otherwise. Boiling is considered the safest purification method, and the longer you boil the water, the safer it is. Don't forget to carry extra stove fuel if you plan on boiling water on your stove.

2) You can use *water purification tablets* which are a lot more convenient but occasionally not as effective as boiling. Halazone (chlorine) tablets are still popular but iodine ones are slightly more effective.

3) Carry a small amount of liquid *household bleach* in a ½-ounce plastic container. Use a plastic eyedropper to place 2 drops per quart of water into your canteen. Rinse out the eyedropper afterwards, since bleach will corrode its rubber top part. While 2 drops are sufficient for most conditions, use up to 4 drops for obviously polluted water.

4) Place 3–5 drops of *tincture of iodine* in each quart of water. Carry and use it as described above for bleach.

After applying tablets, bleach, or iodine, wait 5 minutes, then shake the canteen with the lid loose enough so some water leaks out to cleanse the rim, and wait 20 more minutes before using. Double the times mentioned above when using tablets, bleach, or iodine to purify water colder than about 40°F, and double the indicated dosage if noticeably contaminated water is used. You can mask the chemical taste of purified water with a fruit-flavored powdered drink and improve the taste of boiled water by pouring it back and forth from container to container several times.

Portable *water purification devices* appearing on the market are designed more for camping than for serious backpacking. Few experienced hikers prefer them over the methods described above. Although effective, they are generally expensive and heavy.

If no insects, plants, or other signs of life are in or near a spring or stream, it's probably contaminated with a chemical poison, which no amount of purification will make safe to drink. Chemically poisoned springs are rare and found primarily in the southwestern states.

Clear muddy or dirty water before purifying it by letting it stand still for a while so gravity can pull the mud particles to the bottom, and then slowly pour off the clear water. Another method is to simply pour the muddy water over a filter made by piling a 1- to 2-inch layer of sand over a shirt held above a container.

Snow is almost always safe to use untreated. Snow obtained from large snowfields or in the middle of winter is safest, while snow obtained from the last lingering snowdrift in spring is of more questionable quality.

Melting Snow

There are several ways to *melt snow* into drinking water:

1) Melt snow in a pan over your stove or a fire by filling it three-quarters full of loose snow or ice and adding some water from your canteen to it. When about half the snow melts to water, pour all but 1 inch of that water into another container. Then add more snow or ice to the pan, heat it, and pour it into the extra container when mostly melted.

Repeat the procedure until you have enough water, but always leave a layer of water in the pan when doing this to prevent the snow from getting an awful burned taste.

2) Fill a canteen with loose snow or a snow/water mixture and carry it in an outer pocket of your clothing so your body heat can melt it during the day.

3) Spread some snow on a black plastic sheet in a slight depression in the sun. The sheet will absorb the sun's warmth and melt the snow. For more rapid melting, place a foam pad under the plastic sheet to insulate it from the cold ground.

4) Place a pot of snow on top of another pot of melting snow or food cooking on your stove to better use all the stove's heat (Example 12-11).

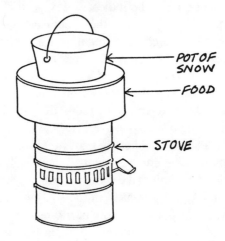

Example 12-11: Using a stove to cook food and melt drinking water.

When carrying large amounts of water from a water source to your campsite with a bucket, line it with a large plastic bag, fill that bag with the water, and seal it shut with a rubber band. This prevents spills which waste water and helps you carry more of it at one time to reduce the number of trips you'll make getting it (Example 12-12).

Hiking Etiquette

Be considerate of other trail users. Horses, mules, downhill hikers, and faster hikers walking in your direction have the right of way. When horses or mules are passing you, stay in full view of them by standing on the outside (the downhill side) of mountain or canyon trails and talk softly to them so you don't spook them. Don't cut across switchbacks because that causes trail erosion. Smoke cigarettes only when resting and only on a rock or nonburnable surface. Field dress the butt by putting its outer paper and filter in your garbage bag. Don't walk on Indian ruins, deface petroglyphs, or remove any artifacts from any place with aesthetic, cultural, or historical value. Respect private property signs, get permission to camp anywhere that's not public land, leave all gates as you found them (open if found open, closed if found closed), obey all fire-building and camping regulations, carry your unburnable trash home with you, and leave the trails and campsites cleaner than before you came there.

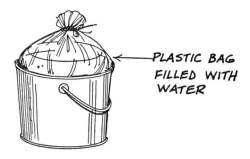

Example 12-12: Carrying water with a plastic bag inside a bucket.

13

Your Campsite

Selecting a Campsite

It's time to retire the pioneers' dig, chop, alter, construct, and destroy mentality for the frontier of the past is gone. Now we must protect the few remaining tracts of undeveloped land in the country by educating the hordes of 20th century outdoorsmen whose axes chop down trees, soap suds float in rivers, and fire pits line the trails. Now all of us must camp without building bough beds, without ditching tents, and without removing a single leaf from our campsites. Now more than ever before, we need to take nothing but pictures, leave nothing but footprints, and kill nothing but time.

You should camp simply without building lean-to's, fancy campsites, or log-lashed tables. Camp so that no one else knows you are or have been there. Don't camp within 200 feet of any water source to keep it unpolluted. Camp below treeline in the mountains to preserve the fragile tundra vegetation. Use old, existing fire pits instead of building new ones, but if you have to build a new one, destroy it when you leave the area. Camp in designated sites in a park or wilderness area to localize the human impact and leave the greater expanse of remaining wilderness untrampled.

There are many extremes of camping. Some people like to establish a campsite with the conveniences, comforts, and organization typical of their city homes and lifestyles. Others are happier if they can simply crawl into their sleeping bag in the thickest part of a remote wilderness. In fact, despite your initial preconceptions of what camping is like or should be, with a little experience you can camp comfortably almost anywhere. For general discussion purposes and more to help beginners get started than to perfect advanced hikers' techniques, the next several paragraphs offer suggestions on how to select and set up a campsite.

A good campsite should be close to *water* for drinking, and should provide adequate *firewood*, splendid *views*, *shelter* from storms and heat, unrestricted *sunshine* for warmth, a soft, *flat surface* for your sleeping bag, a gentle *breeze* to keep bugs away, and *privacy* from other campers but *companionship* with them. Obviously, there is no perfect campsite, so don't waste your time looking for one, but the more of those features a site provides, the better it is for camping.

To avoid bugs, camp on a hillside out of the trees where breezes are more common. Avoid surface water where bugs breed.

To avoid the sun, camp in a forest or on the north side of a hill, fence, grove of trees, or other obstruction.

To avoid the wind, camp in valleys, in a forest, or on the leeward side of a hill, boulder, or cliff.

To avoid the cold, camp on south-facing hillsides to get as much sun as possible and not on mountaintops, where strong breezes constantly blow or in valleys where damp, cold air collects in the evening.

To avoid moisture, camp on hillsides and not in damp valleys or near bodies of water. Camp in a forest so the trees can shelter you from dew or from the beginning of a storm, but camp in an open field or meadow to take advantage of the sun's drying rays after a storm has passed.

To wake up early, camp in a field so the sun's first rays can strike your campsite.

To sleep late, camp behind an obstruction like a hill, a group of trees, or a boulder that blocks the sun's morning rays.

In *deserts* camp away from washes which could flash flood; in *mountains* camp away from steep slopes which could avalanche snow or rocks; and along the *coasts* camp well above the high tide line.

For long-distance backpackers, a campsite is simply a place to sleep until their hike resumes in the morning. Any fairly level place will suffice. For more leisurely backpackers, who enjoy camping as well as hiking, a good campsite has a *sleeping area*, a *cooking area*, and a *bathroom area*, which should be removed from each other and from high traffic areas like paths to a nearby stream or the main trail, if possible. Unless you're experienced in camping in the darkness, allow plenty of time to set up your campsite and cook dinner, especially if cooking on a fire.

Sleeping Area

Set up your tent or tarp away from isolated trees which attract lightning and trees with dead branches that could fall on you in a storm. Set up the shelter on flat or slightly sloping ground for comfort and good drainage and not in a gully or depression which collects water when it rains. Face it away from the wind so an unexpected storm won't blow the rain in on you, and rig it upwind from your campfire to keep smoke and sparks away from it. Be sure you clear the ground of any sharp sticks or rocks before setting up your shelter, especially if it has a floor. Also, it's a good idea to lay on your ground sheet on the ground before setting up your shelter to see if it's a comfortable place to sleep.

Put tent and tarp stakes in the ground at a 90° angle to the direction of the guy rope for maximum support. In soft ground, overlap two stakes vertically, while in snow bury two of them in a horizontal "x" pattern for extra support. If the ground is too hard to push stakes in, tie the guy ropes to trees, bushes, or rocks (Example 13-1).

After you set up your shelter, spread out your ground sheet (if

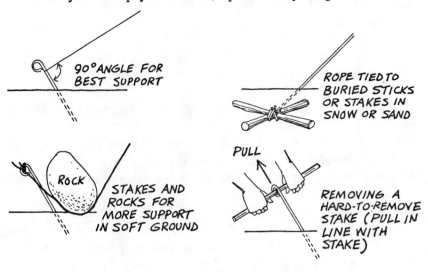

Example 13-1: Using stakes properly.

sleeping in an unfloored shelter) and foam pad and lay on them (with your head slightly uphill for more comfort) to determine if the ground's natural bumps and contours bother you anywhere. If they do, shift to a slightly different position. Then, put extra clothes under your head, the small of your back, and your thighs for a more comfortable sleep (Example 13-2). You can sleep comfortably on very hard surfaces using the padding method described above.

Unroll your sleeping bag at least one hour before using it so its insulation has time to fluff up for greater warmth at night. In damp or wet weather, however, unroll the bag the instant before you use it to keep it as dry as possible. Before going to sleep, place items like your flashlight, extra sweater, alarm clock, canteen, and insect repellent near your sleeping bag so you can reach them if needed. In cold weather place your stove and canteen nearby so you can make a hot drink or cook breakfast in the morning without getting out of your sleeping bag. Put your eyeglasses in a carrying case or slip them into one of your boots for protection. When sleeping on uneven ground, place rocks, logs, extra clothing, or your hiking boots on the downhill side of your sleeping bag to keep from sliding downhill during the night (Example 13-3).

When sleeping without a shelter, point your feet in the direction of the wind so drafts can't blow in the top sleeping bag opening and chill you. For extra protection from the wind, place your feet inside your emptied backpack or drape a windproof parka over them.

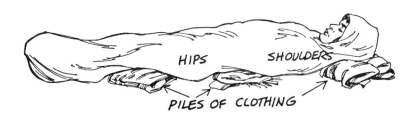

Example 13-2: Pile clothing, leaves, or sand under your head, the middle of your back, and your thighs for more comfort when sleeping.

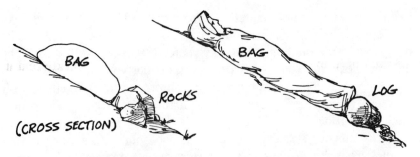

Example 13-3: Pile extra clothes, your boots, rocks, logs, or your pack against your sleeping bag to prevent sliding downhill at night.

Cooking Area

You can cook almost anywhere with a stove as long as you select a place sheltered from the wind and away from any burnable materials like pine needles. Chapter 16 explains how to build a fire and cook on it.

Refrigeration

Although it's often best not to carry foods requiring refrigeration, you can keep perishable foods fresh on a hike by storing them in water-proof containers in a river, lake, or stream. Surround your food cache with rocks if the current is strong. Foods not highly perishable will last for several days in warm weather when wrapped in a bundle of clothing and stored in your pack or in the shade. You risk losing your food and attracting very unwanted dinner guests in certain areas when keeping food in your backpack or a nearby stream, though. See page 260 for more information about storing food properly.

High Altitude Cooking

The boiling point of water is 100°C at sea level but falls 1°C for every rise of 1,000 vertical feet. Thus, water boils at 90°C on a mountain 10,000 feet above sea level. This means that you'll need more time (and thus more water and stove fuel) to cook foods at higher altitudes. The best way to judge how much longer to cook food is to taste it. If it tastes cooked, then it's ready to eat.

Sanitation

Disposing of Trash

Burn your paper, plastic packaging, dirty toilet paper used for clean-ing dishes, leftover food scraps, and the food residue remaining in any cans in a campfire. If you can't or didn't build a fire, burn easily combustible materials like paper in a very small fire and dump any leftover food several hundred yards away from any trail, campsite, or water source. You don't need to bury waste food because animals will quickly find it and eat it even if it's buried. Carry all unburnable garbage like cans, glass, and aluminum foil home with you.

Washing Dishes

It's best to clean dishes immediately after use before leftover food hardens on them. To clean your dishes, first wipe any food residue out with toilet paper and burn that paper in a fire. If you don't have a campfire, light it with a match and burn it. Then use sand, snow, or a scrub pad to scrape away hard-to-remove foods. (When cleaning pots with sand or snow, you can often skip the initial toilet paper wiping.) If water is scarce or if you're in a hurry, you'll have to stop with this step and wash the dishes more thoroughly later. If water is readily available, sterilize your dirty dishes by washing them with soap or by boiling some water in them. Rinse all soap-washed dishes thoroughly, since even a small amount of soap residue in them could give you diarrhea or an upset stomach. Above all, wash dishes at least 200 feet from any water source. Washing dishes with soap and sterilizing them with boiling water is much more important for groups of people. Solo hikers can ignore these steps if they adequately remove all food residue from their eating utensils.

When camping with a group of people, each individual should clean his own utensils using the toilet paper and scrub pad method described above. Then several volunteers or people assigned to clean-up duty should clean all the community cooking pots in the same fashion. At the same time someone else should collect water (preferably previously heated) in three buckets or in three constructed containers using logs or rocks (Example 13-4). It's a good idea to put a bucket of water on your campfire or stove to heat while you eat to speed up this cleanup process.

Then the cleanup people should thoroughly wash all the utensils and dishes in soapy water, rinse them in warm water, and sterilize them in hot

Example 13-4: Constructing containers for washing dishes with logs and a plastic sheet.

rinse water. Replace the water in the first rinse basin as soon as it becomes dirty or slightly soapy. For large amounts of dishes, replace the wash and rinse water several times. Dishes air dry quickly if rinsed in hot water. Otherwise, towel or air dry them and pack them away when dry. When finished, dump the soapy, dirty dish water at least 200 feet from any trail, campsite, or water source.

Washing Yourself

Although you'll get used to feeling dirty after a few days of hiking, it's good to bathe occasionally, especially on longer trips. If the weather is warm and a storm is approaching, use the rain for a shower. Simply remove your clothes, lather with soap, and let the rain wash it away. If it's not raining, construct a small bathtub with logs and your ground sheet as described above, fill it with water, and sit in it if it's large or use it as a sink if it's small. In cold weather, heat some water on your stove and wipe yourself clean with a bandana. Dry yourself off in the sun, with your bandana, or with an extra shirt you can dry out later. Feel free to swim anytime to wash grime and sweat off your skin, but never wash yourself with soap in any river, lake, or stream.

Human Wastes

Human wastes must be disposed of properly to keep campsites, drinking water sources, and the environment clean. Use outhouses wherever available and always urinate and defecate at least 200 feet from any campsite, trail, or water source. If you're hiking alone or in a very small group of people, urinate on soil which filters your wastes and defecate in a shallow hole in soil about 6 inches deep. Natural bacterial action in the soil will decompose your wastes quickly. If the hole you dig is too deep or if the soil is sandy, it'll take longer to decompose, though. Burn the toilet paper you use, since it takes a long time to decompose, but never burn it in places where you could start a forest or a field fire (surprisingly, easily done). When finished, cover the hole so no one knows you were there.

When camping with a larger group of people, select a location for a group latrine that's secluded and removed from the campsite area and is at least 200 feet from any water source. Then dig a trench 6 inches deep and 1–2 feet long and pile the soil you remove to one side so that everyone can use the same trench and can cover their wastes with some soil when done. Consider locating your latrine near a log for a seat or a tree branch to hold onto for support. Keep your toilet paper at the latrine and covered with a plastic bag so no one has to hunt through the campsite for it.

Group Camping

There are several ways to camp with an organized group of people:

1) Everyone knows what chores must be done around the campsite and they instinctively volunteer to do them. This works best with families, close friends, experienced backpackers, and very small groups.

2) Larger groups need an appointed leader who directs the operations of the campsite by assigning everyone specific chores and responsibilities, like firewood collecting or cooking. This works best with groups of kids and people who are not close friends and in situations like summer camps, Boy Scout outings, and professional guide services, where the leader stands out from the rest of the group because of his greater experience or maturity. A rotating schedule of chores assures that everyone including the leader does an equal amount of work (Example 13-5).

	Mon.	*Tues.*	*Wed.*	*Thurs.*	*Fri.*	*Sat.*	*Sun.*
get wood	Tom	John	Sam	Bob	Kurt	Ed	Tim
get water	Tim	Tom	John	Sam	Bob	Kurt	Ed
cook	Ed	Tim	Tom	John	Sam	Bob	Kurt
cook	Kurt	Ed	Tim	Tom	John	Sam	Bob
wash dishes	Bob	Kurt	Ed	Tim	Tom	John	Sam
set up tents	Sam	Bob	Kurt	Ed	Tim	Tom	John
relax	John	Sam	Bob	Kurt	Ed	Tim	Tom

Example 13-5: Organizing a group for camp chores.

Leaving a Campsite

Leave a primitive campsite as if no one knew you were ever there by always replacing everything as you found it. Destroy your firepit if you built one by burying or scattering the dead ashes and returning the firepit rocks to where you found them. Scatter your woodpile, ruffle up the leaves where you slept, and collect all the litter you find in the area. Leave the place as wild as possible. At developed campsites or campsites which you're sure others will use, though, leave the firepit in place and leave any remaining firewood nearby.

Low-Impact Camping

In these times of crowded parks, overused trails, and congested camping areas, it's extremely important that you follow the principles of *low-impact camping*:

1) *Preserve everyone's wilderness experience.* Always camp so that no one knows you are or have been at a campsite. Always leave a place cleaner and more wild than when you found it.

2) *Preserve the immediate resource.* Cardinal sins include camping next to desert springs, camping in fragile alpine meadows, leaving ugly fire pits in remote places, and washing yourself, your clothing, or your dirty dishes in water sources.

3) *Preserve remaining wild areas.* Educate, don't criticize, other hikers ignorant of proper outdoor practices and support conservation organizations (see Appendix for their addresses).

Camp Security

Animals, people, and the natural elements could threaten your personal safety and the security of your gear at your campsite. Since *animals* are primarily interested in your food, you can virtually eliminate that threat with proper food storage (see page 260).

While most *people* are extemely friendly and helpful outdoors, a few are downright nasty. In general, the farther away from highways, communities, and popular parks, trails, and campsites you go, the safer you and your gear will be. Camping with a dog, with a group of other hikers, and in a closed shelter like a tent increases your safety. When camping in popular areas, it's a good idea to keep all valuables hidden all the time. Whenever you leave any campsite unattended, hide valuables like cameras inside your sleeping bag and tightly close your tent zipper to discourage sticky-fingered visitors.

Natural elements like the wind and rain could destroy your campsite if you're not careful. Always securely pitch your shelter. The time when you are too lazy or too tired to properly stake down your tent is the time a gale will blow in at 2 a.m. and tear it to shreds. Never leave a fire unattended, even for a few minutes. It could spread out of control, burn up your gear, and burn down the forest. Be extremely careful with fire on windy days. Never leave gear outside your shelter or drying on a clothesline in unstable weather. A wet down sleeping bag will take several days to dry if soaked in a sudden downpour.

14

Expedition Backpacking

Expedition backpacking is hiking for a great distance or for a long amount of time. Only experienced, well-prepared, physically conditioned hikers should attempt it.

Planning

Since your pack will be the heaviest during the beginning of the trip, plan on hiking the fewest miles per day at that time and the most miles per day towards the end of the trip when most of your food weight is gone and your body's in better condition. Don't overexert yourself during the first few days, either, because you'll be too exhausted for the rest of the hike. Plan the trip to allow at least one or two rest or low-mileage days for each six to ten days of regular hiking to give you time to relax, do laundry, resupply with food, or explore on a day hike for a change of pace. It also allows flexibility in case of bad weather or unexpected delays, so you don't absolutely have to get to a certain campsite or hike a set number of miles each and every day without falling behind in your plans.

On long hikes carry only the equipment you really must have and equipment that has a variety of uses, like a bandana for a towel, dish cloth, and potholder, and socks that can double as gloves for your hands. The longer and harder the trip, the more critical the weight factor becomes and the more selective you must be in choosing your equipment.

Before going on a long backpacking trip, go on several weekend "shakedown hikes" to get in shape, test your equipment, and perfect your plans, menus, and camping techniques. Be sure all your gear is in excellent working condition and you're familiar with all the information written about the area you'll hike in. Pore over topographic maps to check and recheck your water supplies, hiking routes, campsite locations, and food resupply points.

Food

Select foods that are extremely lightweight yet very nutritious and high in energy. Carry plenty of food, since you can't rely on your stored body reserves like you could for a short weekend hike. Carry vitamin pills and eat a balanced diet to assure proper nutrition. You may need to rely entirely on specialty dehydrated foods which are the lightest and least bulky of all hiking foods. At the most, you'll be able to carry about two weeks' worth of food without resupplying in warm weather and only about seven to ten days' worth of food without resupplying in cold weather, although a few hearty hikers can carry more and still enjoy their hike.

On very long hikes, there are six ways to resupply yourself with food and other expendable necessities like toilet paper, soap, and matches:

1) The cheapest method is to plan your hike so that you can *purchase supplies from grocery stores* along your way. Many hikers on the Appalachian Trail use this method, since the trail passes near small towns every several days or so. A disadvantage is the limited selections available at many small-town stores.

2) If you want to eat strictly catalog dehydrated foods and aren't concerned about cost, have a friend *mail your supplies to post offices* along your way. Mail them to: your name, General Delivery, Town post office, Town, State, and zip code. Mark all packages "Hold for through hiker—expected—_____" and give a three week time span to allow for mixups and delays. Be sure you aren't shipping your supplies to a seasonally operated Post Office. Also, since the Post Office is only required to hold general delivery mail for 15 days, it's a good idea to send the Postmaster at your supply point a card requesting that he hold your package if you don't receive it within their time limit.

3) *Burying supply caches* along the way interrupts the continuity of your hike the least because you don't have to plan time to walk to a town,

but you must travel along your route ahead of time to hide your supplies and again afterwards to remove your empty cache containers. Bury the caches in secure, waterproof containers deep enough to keep animals out and camouflage their location so other hikers won't find them. Mark their exact locations on your maps and in a separate notebook so you can find them again. Avoid foods damaged by excessive heat and cold, and always leave a note with each cache saying not to disturb it in case someone accidentally finds one. (Once, on a 110° summer day in a remote part of the Grand Canyon, I drank a six pack of beer that was cached for two years in a cold stream—thanks, Tony!)

4) You can have *another backpacker resupply you* along your hiking route but you need to make exact arrangements for this days ahead of time and that could cause complications if last-minute changes occur.

5) You can *load your car with your supplies* and leave it at the halfway point, although that complicates your transportation situation before and after the hike.

6) A final and very expensive method for wilderness hikers is to have supplies dropped by helicopter, airplane or boat, but that's not very practical for almost all hikers almost all of the time.

Companions

Hiking with other people poses great problems, especially on long-distance hikes. Personal conflicts easily develop over any number of things, like how far and how fast you'll walk, who should do the camp chores, and where you'll camp for the night. On extended outings, severe conflicts even develop among the closest of friends and married couples over the most trivial thing. Because of the problems associated with hiking with a group of people, many expedition hikers walk alone.

The following suggestions will help you prevent a personal problem from ruining your hike:

1) *Know your partners.* Often it's best to be good friends rather than just casual acquaintances before beginning a long hike. Also, everyone should know the purpose of the hike, the group expectations such as how far you plan to walk each day, and each other's strengths and liabilities. A hiking group, like the proverbial chain, is only as strong as its weakest link.

2) Since people occasionally need to be alone and need to be away from the people they are traveling with, *incorporate "personal space" time into each day.* One way to do this is by having everyone sleep in

their own shelter. Another is to have everyone hike at their own pace and meet at a specified destination to camp for the evening. Alternating camp chores so that one person assumes full responsibility for preparing a meal while his partners are free is a third method.

Breaking Free

It will take at least a week of hiking to free yourself from the insidious 20th Century chains of shallow desires, economic pressures, schedules, fast food munchies, and the "it's Friday, we have to party" syndrome that bind and blind us all. Your body needs at least a week of hiking to cleanse its rhythms and thought processes. As you break free, you'll gain a greater perspective of what really matters and what doesn't in life. You'll begin to realize that you don't need a shower every day to feel great, that watching the sunset is more enjoyable than watching TV, that you're healthy living on a steady diet of oatmeal and noodles, and that it's okay to go to sleep when the sun goes down. After you've hiked for long periods of time (especially alone), you'll really understand the meaning of "breaking free."

15

Cold Weather Camping

Because cold is relative—it depends on how well you're prepared for it—the methods and ideas discussed in this chapter apply to people camping in cold weather in any location and at any time of the year. You don't have to be camping at the North Pole to use the information that follows. Use it for any cold weather camping trip, from a winter hike in Florida to a summer backpacking trip in the Rocky Mountains.

Planning and Preparation

If you've never hiked in cold weather, plan a few very easy, undemanding day hikes and backyard backpacking trips first to get used to the special demands it places on you. Even experienced hikers usually plan slow, comfortable trips in cold weather. There's just too much to do to try to cover long distances at a fast pace. You'll need more time to get out of your sleeping bag, take down and set up your camp, pack your gear, and cook. In fact, everything you do becomes very complicated in cold weather. Even a little chore like washing a dirty dish could be a time-consuming hassle when your hands are cold, it's almost dark, you have to use snow to "wash" the pan, and the soup leftovers froze to it. Also, if you're camping in winter, you'll have a lot less available daylight time because the sun will rise late and set early. If the trails are snow-covered, you'll have to walk at a much slower pace—and unfortunately, your pack will weigh considerably more since you have to carry extra food and plenty of warm clothes.

Be completely familiar with all aspects of your equipment. Practice using your stove at home in the dark in the cold so you'll instinctively know how to use it on a trip. Sleep in your tent in a snowstorm in a local park or backyard to see how well it performs in snow and how much insulation you personally need when sleeping. Pack more warm clothes if you sleep "cold" or if you chill easily. Test the layering concept explained on page 82 on a few day hikes or easy overnight trips before relying on it on a challenging cold weather hike. Tie pull tabs on all zippers so you don't have to remove your mittens or gloves to use them (Example 15-1).

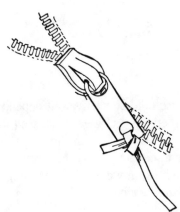

Example 15-1: Tie pull tabs on zippers for easier use in cold weather.

Pre-mix as much of your food as possible into ready-to-eat, single serving bags so you don't have to fumble with three or four plastic bags of separate ingredients with gloved hands in the cold. For example, mix the correct proportions of oatmeal, powdered milk, wheat germ, and brown sugar into individual packets you can just add to a pan of hot water. Carry at least 2–2½ pounds of food per person per day. That should include many high calorie foods like nuts and margarine, because you'll need about twice as many calories when camping in cold weather as you do in warm weather. You'll be warmer if you're never hungry.

Walking in Cold Weather

1) Test ice with a stick to be sure it's strong enough before crossing frozen streams. One step in icy water could be deadly.

2) Walk at a very controlled pace in cold weather to avoid overheating and sweating, which dampens your clothing, draws body heat away by conduction, and eventually chills you. Remove a layer of clothing as soon as you feel overheated.

3) When hiking with a group of people in deep snow, switch the lead often to share the responsibility for breaking the trail. The group can continue walking without stopping by having the lead person step aside every ⅛ to ¼ mile, wait for the others to pass, and assume a position at the end of the line.

4) To keep your socks dry when hiking in wet conditions in cold weather, wear a plastic bag right next to your skin, put one or two pairs of socks on over that bag, put another plastic bag on over those socks, and finally put on your boots. This way your skin will become soaked with perspiration but the sock layers will remain dry and provide maximum insulation (Example 15-2). Seal the top of the plastic bags with a rubber band wrapped around your leg for walking in deep snow or when crossing streams if you have no gaiters. This is similar to the vapor barrier concept described on page 57.

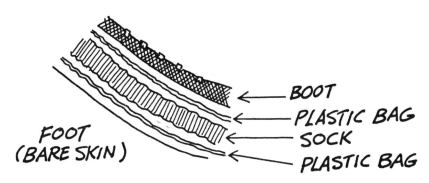

Example 15-2: Using plastic bags to keep your feet dry and warm.

5) Begin hiking at dawn for maximum use of daylight, and so you can walk on top of the snow. As the sun melts the snow during the day, you'll begin to sink into it with each step.

Keeping Warm

Your Feet

1) Put your boots inside a plastic bag in your sleeping bag or wrap them in a jacket and use them for a pillow to keep them from freezing solid in cold weather and to help prevent cold feet when you put them back on in the morning.

2) If your boots are wet, put on a dry pair of socks and walk in them for 30 minutes or so. Then replace them with another dry pair. Repeat this several times until the boots feel almost dry, but don't forget to keep several pairs of completely dry socks in reserve for later use.

3) Because too many pairs of socks restrict the flow of blood in your feet, don't overdress with them in cold weather.

4) If your socks are damp from sweat, turn them inside out for greater insulation.

5) Keep your feet warmer by standing on poor heat conductors like pine needles and logs and not on snow, rocks, or metal objects.

6) If you can't put frozen boots on in the morning, place them in the sun, carefully warm them over your stove, or as a last resort, boil water on your stove and pour that on their outsides to thaw them out. With this last method, they'll be somewhat wet afterwards, but you won't really notice that if you line your feet with plastic bags and immediately begin hiking in them.

Sleeping

1) Air out your sleeping bag each day to keep it as dry as possible.

2) You'll sleep warmer with at least a thin layer of clothing on so your skin doesn't touch against the colder parts of the sleeping bag.

3) Put on fresh, dry socks just before going to sleep. You're asking for cold feet at night if you sleep in the socks you hiked in all day.

4) Put your empty backpack under your feet and put extra clothes between your sleeping bag and the foam pad for insulation from the cold ground. When sleeping in cold weather, you'll lose as much as three-quarters of your body heat by conduction through the ground.

5) Drape extra clothes over your body like a blanket inside the sleeping bag or simply stuff them inside your bag for more insulation.

6) If your sleeping bag doesn't have a shoulder collar, put a piece of clothing around your neck and shoulders to keep cold drafts out.

7) Depending on the wind chill factor (see page 254), sleeping in a tent is about 10–15° warmer and sleeping under a tarp is about 5–10°

warmer than sleeping in the open. Sleeping next to someone else or in a tent with someone else is considerably warmer than sleeping alone, but sleeping with someone in two sleeping bags zipped together is colder than if they are used separately.

8) Eat a lot of food immediately before sleeping. Carbohydrates provide quick, immediate energy while fats provide sustained energy that could last throughout the night.

9) Eat a snack at midnight to refuel your metabolism.

10) Just before going to sleep, warm the inside of your sleeping bag by a campfire.

11) Get dressed and undressed inside your sleeping bag. Although difficult to do, your attempts at it will warm you and the bag considerably.

12) Warm your bag with hot water bottles.

13) Do isometric excercises inside your sleeping bag.

14) You'll sleep warmer at night if you aren't fatigued or chilled from the day's activities.

15) Keep your mouth and nose outside the bag so moisture from your breath doesn't condense inside its insulation. In extremely cold weather, breathe through a sweater draped over your face.

16) When sleeping outside with no shelter, sleep between a small fire and a reflecting object like a boulder, sleep between two small fires, or build a long fire, put it out, and sleep on the warm ground where the fire was (Example 15-3).

17) Sleep on a poor heat conductor like pine needles instead of on cold rocks or damp ground.

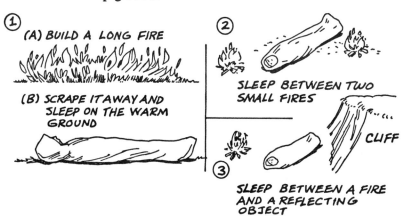

Example 15-3: Using a fire to sleep warmer in cold weather.

18) When sleeping under a tarp or with no shelter, use your backpack or some boulders as a windbreak to prevent unnecessary convective heat loss. Even a slight breeze blowing across your sleeping bag will make it seem 5–10° colder.

19) Use a wide plastic ground sheet to keep as much snow away from your sleeping bag as possible when sleeping in the open or in an unfloored shelter.

20) Keep your feet warm by wrapping them in a sweater, putting them in your sleeping bag stuff sack, and pulling that up to your knees inside your sleeping bag.

21) When camping without a shelter in cold weather, reduce your radiated heat loss by sleeping under a tree or rocky overhang and not in an open meadow.

22) Since cold air settles in valleys and basins and damp air collects near bodies of water, you'll sleep warmer in a forest on a hillside.

23) Before setting up your shelter, pack snow down with cross-country skiis, snowshoes, or hiking boots instead of scraping it away, because it'll provide some insulation under your body at night.

24) As soon as you get out of your sleeping bag in the morning, squeeze all the warmed air out of it so the moisture it contains won't condense in the insulation. Then hang the bag in the sun while you pack up the rest of your gear.

25) Place the stove, water, and some food within reach inside or just outside your shelter so you can cook breakfast without getting out of your sleeping bag. In weather below freezing, keep your water bottles in your sleeping bag so they won't freeze.

26) Sleep inside a vapor barrier and a radiant heat barrier. Both are necessities in extremely cold weather.

27) Place pot-sized blocks of snow just outside your shelter so you can melt water without getting out of your sleeping bag.

General

1) When cold, eat sugars and carbohydrates for quick energy. Eat fatty foods to sustain your warmth before getting chilled, since they require a longer time for digestion.

2) Keep your body moving. Jump, exercise, walk around, stomp your feet, and slap your hands.

3) Cold feet are a sign that your body is becoming chilled. Put on a hat and extra clothes for more insulation.

4) Make faces to increase the circulation to your face and to prevent *frostbite* (the actual freezing of exposed skin).

5) If your fingers are cold, curl them up in a ball inside your mittens or warm them in your armpits. Take your arms out of their sleeves and keep them close to your body if they're cold.

6) Make all your physical actions deliberate and slow to avoid overexerting and perspiring.

7) Air out your clothes in the sun whenever possible to reduce their moisture content. The greater the invisible body moisture in them, the faster they'll conduct your heat away.

8) Remove a layer or two of clothing before exercising and put on an extra layer before becoming chilled.

9) Put rubber bands around the bottom of your pants legs to keep warm air in and cold drafts out. Seal other clothing openings by tucking in shirttails and buttoning collars.

10) Heat is quickly lost at the neck, wrist, ankle, knee, and elbow areas because blood vessels are close to the skin at those places. Always keep them covered in cold weather.

11) Don't kneel or sit in snow or on cold objects.

12) If you don't have a hat, wear something (anything) on your head for insulation. Even a bandana offers some additional protection from the cold. You'll lose up to 75% of your body heat through an uncovered head in cold weather.

13) Eat high energy foods throughout the day. Put peanut butter in your oatmeal and butter in your coffee.

14) Warm water on a fire or stove before drinking it.

15) Avoid cigarettes and alcohol. Smoking constricts the capillaries at the skin's surface and makes you feel colder, while alcohol dilates them and increases your heat loss. That warm glow from alcohol is really a loss of body heat at the surface of the skin.

16) Dress in layers for more insulation and wear only synthetic or wool clothes next to your skin. Avoid all cotton clothing except underwear shorts which prevent chafing. Even a thin cotton t-shirt under your insulation layers is useless for warmth in cold weather, and actually cools your body more than if you didn't wear it.

17) Select a campsite with a southern exposure for warmth and one with an eastern exposure for sun early in the morning.

18) Wear rubber dishwashing gloves inside your regular gloves or mittens (the vapor barrier concept . . . again).

Odds and Ends

1) Make *deadmen stakes* for your shelter by tying your tent guy lines to logs and burying them in the snow. Since they could freeze solid at night and are sometimes hard to dig up in the morning, tie the knot above the snow so you can simply untie it and pull the rope free in the morning without digging up the log (Example 15-4).

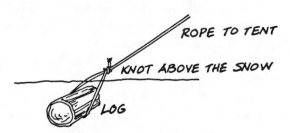

Example 15-4: A deadman tent stake.

2) Flashlight batteries work best when warm, so store them in your parka pockets or sleeping bag at night. Nickel/cadmium or lithium batteries last much longer than regular ones in cold weather.

3) Dry wet clothes by letting them freeze solid and then beating the ice crystals out of them.

4) Put your stove in a plastic bag in your sleeping bag overnight for easier starting in the morning.

5) When cooking, set the stove on an ensolite foam pad to insulate it from the cold ground. Shield it from the wind with rocks, clothes, or your pack.

6) Insulate warmed food by setting dishes on a foam pad. Drink from plastic and not metal cups.

7) Wrap your canteen in a shirt and use it as a pillow or put it in your sleeping bag to keep from freezing at night. Keep it upside down so that if ice forms inside it, it won't freeze the opening shut.

8) Frozen metal water bottles can be thawed on a stove or near a fire but plastic ones can't. Metal ones also resist cracking when frozen better than plastic ones do.

9) When building a campfire in snow, clear the snow from your fireplace or build it on a pile of rocks on top of the snow. If you don't do that, the fire will gradually sink into the snow, melt it, and put itself out. Don't build your fire under a tree branch with snow on it, because the snow could begin to melt or could fall down and smother the fire.

10) Cooking food in winter must always be fast and simple. Save fancy, gourmet meals for warm weather.

11) Camp near open surface water whenever possible. Melting snow for water is a real hassle.

12) Avoid camping in avalanche areas.

13) Daylight time is so valuable in winter that you should constructively use every bit of it. Try to wake up before or just at dawn to cook breakfast and pack up your gear. Also, try to have your campsite completely set up and all camp chores like cooking and melting snow completed before dark.

14) Carry a small *wisk broom* to brush snow from your clothes and gear before entering your tent.

III

CAMPCRAFT SKILLS

16

Fires

In most parts of America, the traditional pioneer campfire is now history. Because of overuse, no firewood remains near many popular campsites like the lean-tos along the Appalachian Trail. Wood in places like the desert is so scarce that it's needed more to nourish the rocky soil when it decomposes than to heat a camper's meal. Often dead wood at high elevations near timberline is several hundred to several thousand years old and is impossible to replace. Don't feel like you have a right to light a campfire whenever and wherever you camp. Only build one if there's enough wood nearby and if it's legal to do so, or if you really need a fire for survival. Although almost all modern backpackers cook on a stove rather than a fire, you should know how to build one in case of an emergency when it could save your life.

General Principles

Selecting a Site

Select a site for your fire sheltered from the wind and away from bushes and trees. Check the wind direction to be sure the fire's smoke and sparks won't blow through your campsite or into your tent. Use an existing firepit if one is nearby. If not, build your fire up against a boulder or encircle it with loose rocks to shield it from the wind, but don't use rocks from a stream or lake because they often contain moisture which could explode when heated. Always remove leaves and other burnable materials for at least 5 feet in all directions from your fireplace and never build a fire on exposed plant roots. Keep a container of water or a pile of dirt nearby to extinguish the fire in case it begins to spread out of control.

Gathering Wood

After selecting a site, gather three large piles of wood. Collect three times as much as you think you'll need, and six times as much as you think you'll need if the wood is wet. Break and pile the wood into three groups—*tinder, kindling,* and *fuel* (see below, and Example 16-1). Put the tinder inside a plastic bag and cover the rest of the wood with a tarp or plastic sheet to keep it dry in case it rains.

Good firewood snaps in half when you break it. If it bends it's still too green, and if it crumbles it's too rotten. There are too many 20th century Daniel Boones running around with axes, saws, and hatchets destroying living trees. Leave those tools (toys) at home and gather wood from the ground with your hands. Leave all standing trees alone. As long as the wood is not too thick, you can break it by bending it on your knee, by placing one end on the ground, holding the other end, and slamming your foot into it, or by propping one end up against a rock and jumping on it. If those methods don't work, simply stick one end in the fire and push it farther in as it burns down. Instead of cutting a long log in half with an axe, burn it in half by placing its middle directly in a fire. Then push both ends into the fire every so often as they burn down (Example 16-2).

Tinder is the dry, flammable material you use to start a fire. Examples of tinder are the cotton fuzz in your pockets, gauze from your first aid kit, dried grass, pine needles, paper, binder's twine, birchbark peeled from dead and downed trees (it flames even when wet), tiny twigs, and

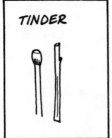

Example 16-1: Kinds of firewood.

BREAK IT ON
YOUR KNEE

SLAM ONE FOOT
INTO IT

BURN THE END IN A FIRE
PUSH IT IN AS IT BURNS DOWN

BURN A LONG LOG IN HALF
THEN PUSH THE PIECES IN

JUMP ON IT

Example 16-2: Several ways to break firewood.

squawwood which is the dead twigs on the bottoms of evergreen trees. Only use squawwood in an emergency to avoid leaving ugly, scarred trees standing. As a general rule, tinder should be thinner than a matchstick. *Kindling* is wood no thicker than your fingers used to ignite the fuel. *Fuel* is wood at least as thick as your arms that keeps the fire burning for a long time. Hardwoods like oak, maple, and birch burn a lot longer than softwoods like pine.

Building a Fire

There are many ways to construct a fire (Example 16-3):

1) To build a *tepee fire*, stick a 12-inch piece of kindling in the ground at an angle facing into the wind. Then place a big pile of tinder under that stick and lean kindling up against it to form a tepee shape around the tinder. Be sure to leave a place open so you can reach your hand in to light the tinder.

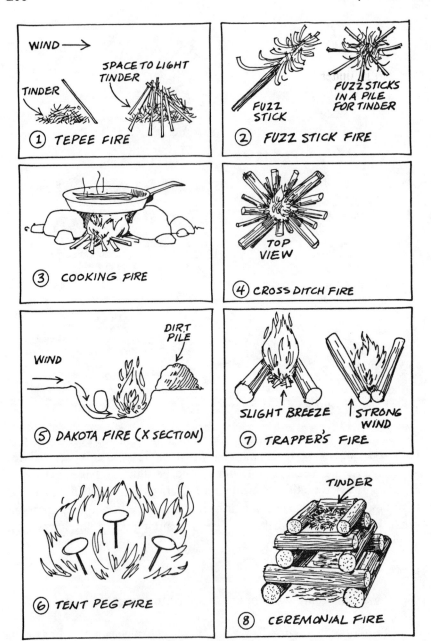

Example 16-3: Kinds of fires.

2) Use your knife to shave thin slices on sticks of kindling, but leave the shavings attached to them instead of cutting them completely off. Make five or six of these *fuzz sticks*, place them in a tight pile, and use them as tinder to start a fire.

3) Build a *cooking fire* by placing two or three rocks on the ground so that they'll form a stable platform to hold your cooking pans. Then place a pile of tinder between them and kindling over that.

4) Make a *cross-ditch fire* by digging an "x" 3 inches deep in the ground and building your fire over it so the ditch can act as a channel to provide air for the fire. Never dig this in rich, organic soil because the soil could smolder unnoticed underground for a long time after you've extinguished the surface fire.

5) Build a *Dakota fire* in areas with severe winds but not in places with soil high in organic matter. Replace the dirt when the fire burns out.

6) Make a *tent peg fire* when you want extra support for your cooking pans. Stick long metal tent pegs into the ground and build your fire over and around them.

7) Build a *trapper's fire* with logs angled in an open "v" facing into the wind for more ventilation in slight breezes. In strong, gusty winds reverse the "v" so that its narrow end points into the wind and shields the fire from it. Also, position the logs so that you can set your pans on them for cooking.

8) Build a *ceremonial fire* by constructing an alternating stack of logs log-cabin style. Place the largest logs on the bottom and smaller ones on top. Angle the logs in towards the center as you go so that by the time you get to the top of the stack, there is virtually no hole left down the center between all the logs. Place your tinder at the top of the pile so that this fire gradually burns down through the pile. This is an excellent method to use when constructing a group campfire.

Lighting a Fire

Light a fire by holding a match in one position under the tinder to concentrate as much heat in one small place for as long as possible. Light the fire from the windward side, and carefully nurse the flame along by placing (not throwing) pieces of tinder and kindling on it. Gradually add thicker pieces of wood as the flames become larger, but don't put too much on too fast or you'll smother it.

A small fire is safer, warmer, and a lot less trouble to keep going than a roaring bonfire. Also, in very cold weather, you'll be warmer sitting between two small fires than next to one large one.

Extinguishing a Fire

Never leave a fire burning unattended. Put your fire out before leaving your campsite and before going to sleep at night. Flood it with water or snow until the soggy ashes are cool enough to touch with your bare hands. If water is scarce, let the fire burn down as much as possible before putting it out. Then slowly sprinkle water on it a little at a time, stir the fire, and sprinkle more water on it. Test it with your hands to see if it's cool and out. If you have no water to use for extinguishing a fire (for example if you're camping in the desert where you need every drop of water for drinking), let the fire burn itself out. Then, to be sure it's out, stamp on it with your boots, pour your dirty dish water on it, urinate on it, or sprinkle sand on it. Again, test the ashes with your hands to be sure they're cool.

Wet Weather Firestarting

Building a fire in a storm is possible to do, although only fools will tell you it's easy. One raindrop can ruin a match and put a butane lighter out of commission if its flint gets wet, so always keep them in a secure, dry place. If they get wet, you'll never be able to start a fire in bad weather.

Before you even think of lighting a fire, though, you need to gather huge piles of wood—at least six times as much as you think you'll need. Look for dry wood under evergreen trees, under logs, on the leeward side of boulders, and in other places sheltered from the rain, but don't just collect dry wood. Gather armfuls of wood, even if it's sopping wet. If you find dry wood, keep it dry under your pack, wrapped in your ground sheet, down your shirt, or in your pockets and keep your pile of wet wood from getting wetter by covering it with your pack, a tarp, or a sheet of plastic.

After you've gathered a huge pile of wood and protected the precious little dry wood (if any) you could find, you begin the slow process of turning wet wood into dry wood. Sit under a tarp out of the rain if possible or stoop over near your wood pile so your back acts like a windbreak and keeps the rain off your lap. Then get some tinder—one piece at a time—and scrape off the wet bark on it with your knife (always carry a knife and matches or a lighter; you'll never be able to start a fire in a storm without them). If the smallest pieces of tinder are soaked completely through, discard them and scrape the wet outer layers off slightly larger pieces of wood. In other words, you may need to begin this procedure with pencil-sized sticks instead of matchstick-sized twigs. After all the wet layers are removed you may have whittled a pencil-sized stick down into a toothpick, but it'll be a dry toothpick. Quickly, before even one raindrop hits it, put it in your pocket, in a plastic bag, or down your socks. Keep it as dry as possible.

Instead of beginning with very small sticks, you can start with a log about the diameter of your arm. Slice off and discard all the wet outer layers of wood and then use your knife to make shavings from the remaining dry inner piece. Save those shavings, keep them dry, and use them for tinder for your fire. If you have an axe (and I assume you don't) you can speed this whole process up considerably by splitting large logs in half and then shaving their insides, which will always be dry, even in a monsoon storm.

You'll need a huge pile of dry tinder and kindling with its wet outer layers removed before building a fire, so don't even consider building one or think about lighting it until you have a lot of dry wood shavings. You can't afford to make any mistakes when building a fire in wet weather. It could take several hours to collect enough dry wood shavings to light a fire, so don't blow it by assuming you have enough when you really don't. When your shavings are finally ready, carefully plan how you're going to build the fire before exposing them to the elements. Generally, it's best to build your fire on a dry platform of logs with their wet layers removed or on the dry undersides of overturned rocks and not on the wet ground.

When everything's ready, carefully light the shavings, guard your tiny flame from stray raindrops, and patiently coax it to life by blowing on it to give it more oxygen. Blow parallel and low to the ground in a slow, steady manner, not down on the flame from above or with a quick gust of wind. If you have a firm, flat object like a plate or a hat visor, swing it back and forth in a regular, controlled motion to fan the fire after it gets going. When the fire is big enough to burn without your constant attention—and fires on cold, wet days should be larger than normal—put wet logs around it or on a grill over it so they can dry before you add them to the fire. When building a fire in wet conditions, you can relax only when the fire has no chance of going out—and that could be hours after you started building it.

I've just described how to build a fire in wet weather when you're alone. If other people are with you, the task becomes easier and less time-consuming because some people can gather wood while others hold a tarp over the place where still others make the dry shavings.

17

Navigation

Maps

A *topographic map* is the kind of map you'll need for camping because it shows natural features like landforms, vegetation, and elevation changes as well as roads, cities, and other works of man in very great detail.

Obtaining Maps

You can obtain topographic maps for your region from a local sporting goods store or from the United States Geological Survey at the addresses below. If you write to the USGS, request a "Topographical Map Index Circular" of the state(s) you're interested in, which tells you which specific maps in that state to order, the booklet called "Topographic Maps" which describes how to use a topographic map, and the sheet titled "Topographic Map Symbols" which describes the symbols used on topographic maps. While each topo map you buy costs about $2, all the other materials are free.

For areas *east* of the Mississippi River, write to:

Distribution Section
United States Geological Survey
120 South Eads Street
Arlington, VA 22202

For areas *west* of the Mississippi River, write to:

Distribution Section
United States Geological Survey
Box 25286
Denver Federal Center
Denver, CO 80225

For Canadian topographic maps, contact:

Map Distribution Office
Department of Mines and Resources
615 Booth Street
Ottawa, ONT Canada KIA 0E9

If you want maps for specific national forests or national parks, write to these addresses:

US Forest Service	US National Park Service
Dept. of Agriculture	Dept. of the Interior
Washington, DC 20250	Washington, DC 20240

You can also obtain maps suitable for camping from your Chamber of Commerce, state department of conservation, or travel club, but topographic maps are the best for backpacking.

Information on a Map

A map's *scale* shows how the distance on the map relates to the actual distance on the ground. Three different scales are generally used for topographic maps: 1:24,000, 1:62,000, and 1:250,000 (Example 17-1).

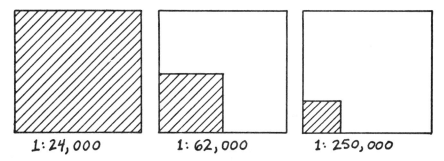

Example 17-1: A comparison of map scales (the shaded region is common to all maps).

1:24,000 means that one inch on the map equals 24,000 inches on the ground. In other words, approximately 2½ inches on the map equals a mile on the ground. Of the three commonly used scales, these maps cover the smallest area but give the most detail for that area. They are best for detailed explorations of a specific area and not for long distance hiking there.

On maps with a scale of *1:62,000*, 1 inch equals 62,000 inches on the ground, or 1 inch on the map equals almost exactly 1 mile on the ground. Maps with this scale are best for longer hiking trips because they cover about four times the area as a 1:24,000 map. Because they only have one-quarter the detail though, many things useful to know for camping, like springs and infrequently used trails, are often not marked on them.

One inch on *1:250,000* maps is almost exactly 4 miles on the ground. These maps are much better for long distance hiking and bike touring than for local hiking because they cover such a very large area that they provide relatively few details for that area. Although maps with this scale give you a general idea of a region's features, you'll probably need to use other maps to obtain more detailed information about that area when backpacking there.

Always check the *date* at the bottom of the map to determine how recent and thus how accurate it is. The map is correct for the year shown on it, but many changes could have occurred since it was printed.

Maps are usually *named* for a significant feature on them, and this name appears in big print at the top of the map. That's the name you use when ordering it. In smaller print and in parentheses around the sides of the map are the names of the maps that border it.

The *small numbers* at the top and bottom of a map are longitude degree measurements, while the ones along both sides are degrees of latitude. They explain exactly which part of the earth the map shows, but you'll never need that information while camping.

Map symbols are colors and shapes that designate specific kinds of landmarks. There are five major kinds of symbols:

1) *Man-made and cultural features* including roads, houses, railroads, powerlines, boundaries, and towns are printed in black or red.

2) *Water features* including rivers, lakes, oceans, swamps, and springs are printed in blue.

3) *Vegetation features* like forests, marshes, and scrub thickets are printed in shades of green.

4) *Geologic features* like contour elevation lines are printed in brown.

5) A purple color shows *revisions* based on aerial photography since the map was first printed.

For a complete listing and description of symbols used on topographic maps, write for the free USGS booklets mentioned previously on page 210.

Contour lines are imaginary lines connecting every point on the map which are at the same height above sea level. The *contour interval* is the distance in *vertical feet* between contour lines, and is marked at the bottom center of all topographic maps. Maps for flat areas like the Great Plains have contour intervals as small as 10 feet, while maps for mountainous areas like the Rockies have contour intervals of 100 or 200 feet.

You'll understand the concept of contour lines better if you imagine a pile of 1-inch thick books on a table with the smallest on top. In this simplified case, (Example 17-2) the contour interval is 1 inch. The books actually look like Figure 1 but are drawn like Figure 2 on a contour map of them. The contour lines simply represent a top view of the pile of books. Where the lines are close together, the books form a steep edge, while they represent a flatter plain where they are farther apart. In other words, the closer contour lines are to each other, the steeper the land is, while the farther apart they are, the flatter the land is. Contour lines that touch each other denote a cliff.

When traveling cross country, use a map with the smallest possible scale and contour interval to give you the most detail. Fifty-foot cliffs

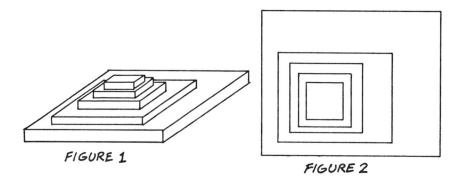

FIGURE 1

FIGURE 2

Example 17-2: Figure 1: A pile of books as seen by an observer. Figure 2: A contour map of those books.

won't show up at all on a map with an 80-foot contour interval (Example 17-3).

Because of its inherent bends and up-and-down sections, it's often difficult to measure the actual distance along a trail on a map. However, you can estimate the trail distance on a map by following the trail with a piece of string and then comparing the length of the string with the map's scale. Add 10% for a straight trail on flat terrain and as much as 40% more for a winding trail in hilly terrain to account for changes in elevation.

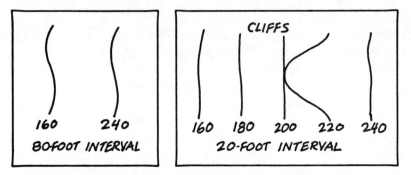

Example 17-3: A comparison of an 80-foot and a 20-foot contour interval (same location).

Orienting a Map

To orient a map, simply line up north on the map with north on the ground by turning the north side of the map so it points north. That means that if you're walking in a southerly direction, you need to look at the map upside down to do this properly, while if you're walking east or west, you need to read it sideways. Usually the top of the map points north, but check the small diagram at the bottom of the map that indicates "true north," "TN," or an arrow pointing north to be sure (see Example 17-8, page 221). When orienting a map you do not need to distinguish between *true north* and *magnetic north* (both described on page 220) since the angle between them is usually relatively small.

Protection and Storage

Using maps covered with clear plastic contact paper or a special plastic waterproof paint sold in otudoor stores is less trouble than keeping them in waterproof plastic map cases while camping in inclement weather. Protect them rolled in cardboard tubes or laid flat and unfolded at home when not in use. Because folded maps wear out quickly, use them unfolded whenever possible, although that's difficult to do when camping. It's a good idea to write trail data and hiking routes on maps with a permanent ink pen so you'll have that information at your fingertips for later use.

Compass

Kinds of Compasses

Fixed dial compasses have a magnetic needle protected by an immobile outer covering, and range in quality from the inexpensive variety found in gumball machines to much better army surplus models. In use they're relatively inaccurate because they have no sighting mechanism to line up with the magnetic needle to indicate a very specific direction.

Floating dial compasses house the magnetic needle in a dial that pivots or turns. They usually have a "direction-of-travel arrow" sighting mechanism for more accurate measurements.

Orienteering compasses (Example 17-4) are floating dial compasses attached to a clear plastic baseplate designed to be used with a map. As you read the next several pages, I assume you have this kind of compass since it is the most useful and versatile for backpacking. Typically, a "direction-of-travel arrow" is the sighting mechanism used on these compasses.

Sighting compasses (Example 17-3a) give you extremely accurate

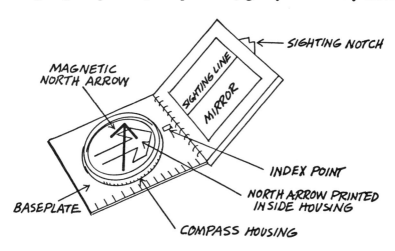

Example 17-3a: A sighting compass with a mirror-index-point sighting mechanism.

readings. This kind of compass is needed when off-trail hiking almost anywhere and when either on- or off-trail hiking in remote areas or under severe conditions. Its added reliability could save your life in an emergency. Some hikers carry several sighting compasses to verify readings on extremely remote, challenging, off-trail hikes. For example, Eric Ryback carried three compasses on his solo Canada-to-Mexico Pacific Crest Trail hike when many sections of that trail were still in the preliminary design stage. Directions for using a sighting compass with a mirror-index-point sighting system are given on page 219.

Better quality compasses are *liquid filled* to slow the needle down so you can read it faster, have a *needle lock* to prevent needle bearing damage caused by shock, automatically adjust the magnetic *declination* (see page 221) for you, and often are designed with special mirrors, dials, and gadgets to increase their accuracy. For general backpacking use and in all but the most exacting circumstances, a general $10–20 orienteering compass similar to the one in Example 17-4 will suffice.

Compass Principles

Compasses are based on mathematical principles, which you should understand before using one. People who study geometry tell us that a

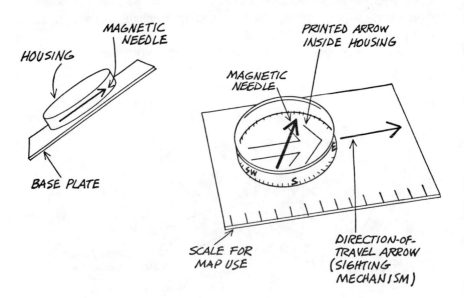

Example 17-4: Features on an orienteering compass.

circle is divided into 360 parts called *degrees*. Thus, half a circle has 180 degrees (180°) and one-quarter of a circle has 90 degrees (90°). The small circle placed after the numbers is the symbol for degrees. We label the most important directions on a compass with direction names as well as with degrees. 0° means north, 90° means east, 315° means northwest, and so on. Remember that 0° and 360° are really two names for the same place on the circle. Since a compass is just a tool to measure a circle, when you use a compass to measure a direction, you're measuring the degrees of direction from a standard point which is 0° or north to another point (Example 17-5).

General Use

When using a compass hold it level and away from all metal objects for the most accuracy. On a hike, a mistake of 1 degree will lead you 92 feet off course for every mile you walk, and you can end up as much as ½ mile off course with a 5° error on a 5-mile hike. Never store a compass near any electrical power supply sources or with other compasses or magnetic items which could ruin its magnetic attraction.

Throughout the sections that follow, a heavy, dark arrow represents the magnetic north needle, while a light, hollow arrow stands for the arrow printed inside the compass housing. Refer to Example 17-4 for terms you are not familiar with.

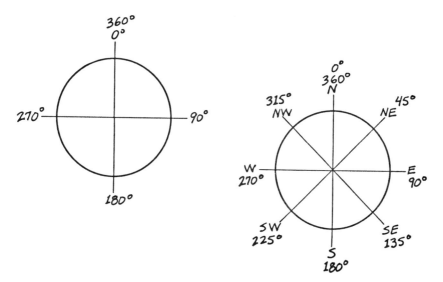

Example 17-5: Compass degrees and directions.

Orienting a Compass

Follow these steps to *orient a compass*:

1) Hold the compass level.
2) Hold the baseplate still while turning the compass housing until the printed needle on the bottom of the housing lines up with the magnetic north needle.

STEP 2:

Taking a Bearing

Taking a bearing means to measure the direction from where you are to a distant object or to your destination. Follow these steps to take a bearing:

1) Hold the compass level.
2) Point the direction-of-travel arrow on the baseplate at the object or in the direction you want to go.
3) Orient the compass (see above).
4) Read the number in degrees from the side of the compass housing indicated by the direction of travel arrow. That number is the *bearing* from where you are to your destination or to the object.

DESTINATION

270

STEPS 3, 4:

Following a Bearing

To *follow a bearing* means to walk in a straight line in the direction of a given bearing. To follow a bearing:

1) Hold the compass level with the direction-of-travel arrow pointed away from you.

2) Turn the compass housing until the bearing you are supposed to follow lines up with the direction-of-travel arrow.

270

STEP 2:

3) Without changing any compass setting, slowly spin your entire body around until the printed arrow on the bottom of the compass housing lines up with the magnetic north arrow.

4) Now the direction-of-travel arrow is pointing in the direction you want to go. See page 275 for ways to walk in a straight line on that bearing.

STEP 3:

Using a Sighting Compass

In general, you use a compass with a mirror-index-point sighting device exactly as you would an orienteering compass, as explained on the preceding pages. The only difference is using the mirror and index point instead of the direction-of-travel arrow as your sighting mechanism. As an example of this difference, the following steps explain how to take a bearing with this kind of compass. Refer to Example 17-3a on page 215 for terms you aren't familiar with.

1) Hold the compass flat, at eye level, and with the hinged cover pointing at an object you want to take a bearing of.

2) Fold the mirror down so that you see the compass dial reflected in it.

3) Using only one eye, line up the object with the sighting notch and the sighting line with the reflection of the index point on the mirror.

4) Without moving your eye or the compass, twist the compass housing until the printed arrow inside the housing lines up with the arrow pointing to magnetic north. You should see this in the mirror. The bearing to that object is indicated and preserved by the index point on the baseplate.

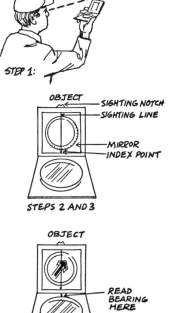

Compass and Map Together

Magnetic and True North

Navigation becomes somewhat more complicated when you use a map and compass together. The whole earth is a gigantic magnet with a north and south pole, but unfortunately the magnetic north and the true north poles don't coincide. The magnetic north pole lies somewhere in northern Canada about 1,400 miles south of the true north pole. In North America magnetic north is up to 25° east of true north in Oregon and 20° west of true north in Maine (Example 17-6). The only places in the continental United States where a compass points directly north is along a line running from Illinois through Georgia. The small drawing (Example 17-7) at the bottom of a topographic map shows true north, magnetic north, and the difference in degrees between them for that region.

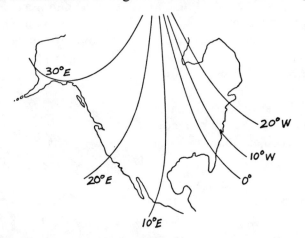

Example 17-6: Differences in magnetic north at different locations (E means east declination, W means west declination).

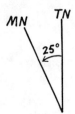

Example 17-7: The diagram indicating true north (TN), magnetic north (MN), and the difference between them in degrees.

Three Methods of Use

The three major ways to use a map and compass together are described below. Practical problems explaining how to use the first two methods then follow.

Method #1—Work only with magnetic north: With this simple method you line up magnetic north on the map with the compass magnetic north needle. What matters is that the map and the compass "speak the same language," which in this case is magnetic north, and not that magnetic north is almost always not the same as true north.

Method #2—Draw magnetic north lines on the map: Drawing magnetic north lines on your map and using your compass as a protractor on these lines is similar in principle, though harder to initially understand, than the first method. Simply place a yardstick along the drawing indicating magnetic north at the bottom of your map (Example 17-7) and continue that line all the way across it. Then make other lines about 1 inch apart and parallel to the first one across it until a grid pattern covers the entire map (Example 17-8). With this method you ignore the magnetic north needle on the compass until you remove the compass from the map and use both separately.

Method #3—Adjust for the declination: This method involves adjusting for the difference between true north and magnetic north. It is by far the most complicated method and should only be used when you must walk in a *true direction* like true north or true east. Though surveyors commonly use this method, backpackers will seldom if ever need it. If the following explanation of it becomes too complicated for you, simply skip it and learn either method #1 or method #2 instead.

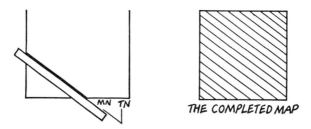

THE COMPLETED MAP

Example 17-8: Drawing magnetic north lines on your map.

Let's say you live in an area with a 15° *west declination* (Example 17-9). That means that magnetic north lies 15° to the west (or to the left) of true north. When going *from a compass on a map to the field*, simply add the declination to the bearing the compass reads. If the compass reads

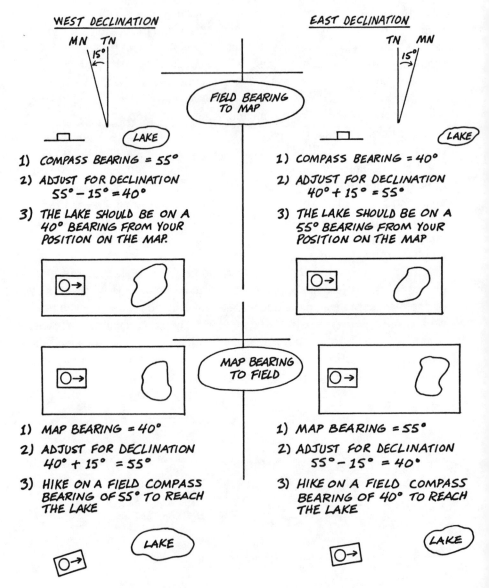

WEST DECLINATION

MN TN

EAST DECLINATION

TN MN

FIELD BEARING TO MAP

LAKE

LAKE

1) COMPASS BEARING = 55°

2) ADJUST FOR DECLINATION
55° − 15° = 40°

3) THE LAKE SHOULD BE ON A 40° BEARING FROM YOUR POSITION ON THE MAP.

1) COMPASS BEARING = 40°

2) ADJUST FOR DECLINATION
40° + 15° = 55°

3) THE LAKE SHOULD BE ON A 55° BEARING FROM YOUR POSITION ON THE MAP

MAP BEARING TO FIELD

1) MAP BEARING = 40°

2) ADJUST FOR DECLINATION
40° + 15° = 55°

3) HIKE ON A FIELD COMPASS BEARING OF 55° TO REACH THE LAKE

1) MAP BEARING = 55°

2) ADJUST FOR DECLINATION
55° − 15° = 40°

3) HIKE ON A FIELD COMPASS BEARING OF 40° TO REACH THE LAKE

LAKE

LAKE

Example 17-9: Comparing east and west declination.

a bearing of 40° when it's resting on your map, just add 15° to it, set your compass to a bearing of 55°, and follow that bearing to your destination when hiking outdoors.

In the same example with a west declination, when working *from a field bearing with a compass alone to a map*, you must subtract (do the opposite of add, because now you're working backwards from the preceding paragraph) the declination from the compass bearing so that the field bearing coincides with true north on the map. If you measured a bearing of 55° to a lake with your compass and wanted to check a map to see if it was the lake you were headed for, subtract the declination (55−15=40) and line up the compass with a bearing of 40° on the map. If the lake is on that bearing, it's the one you were hiking to. In summary, when adjusting a map bearing to a field compass bearing with a west declination, add the declination to the bearing. When changing a field compass bearing to a map bearing, subtract the declination.

When working in an area with an *east declination*, magnetic north is to the east (to the right on the drawing at the bottom of the map). When *adjusting a map bearing to a compass bearing*, subtract the declination from the bearing. When changing a *field compass bearing to a map bearing*, add the declination. The procedures for adjusting an east declination are exactly opposite those for adjusting a west declination. Confused? See Example 17-9.

Three Practical Problems

Problem #1: You want to hike through a forest to an isolated fishing lake a few miles from your camp. You need to use a map and compass to find a bearing from where you are to the lake. In this example, the positions of the lake and your camp are marked on your map.

Using Method #1:

1) Lay the map flat on the ground. Then turn it until its magnetic north line lines up with the compass magnetic north arrow.

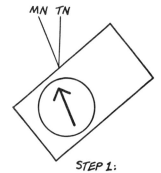

MN TN

STEP 1:

2) Find your position and the lake on the map.

3) Place one of the long sides of the compass baseplate on the map so that it touches your position and the lake. If the compass is not large enough to touch both points, draw a straight line between them with a ruler and place the baseplate on that line. Also, be sure the direction-of-travel arrow points in the direction you want to go. While doing this step, you may need to turn the baseplate while holding the compass housing and map immobile in order to keep the compass magnetic north needle and the map's magnetic north line coincidental.

4) Orient the compass by turning the compass housing as described on page 218.

5) Read the degree measurement where the direction-of-travel arrow touches the compass housing. That is the bearing you must follow to get to the lake (see Following a Bearing, page 218).

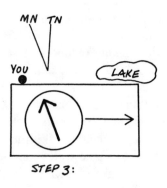

STEP 3:

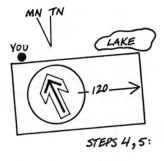

STEPS 4, 5:

Using Method #2:

1) Lay the map flat on the ground with no particular orientation.

2) Find your position and the lake on the map.

3) Place one of the long sides of the compass baseplate on the map so that it touches your position and the lake. If the compass is not large enough to touch both points, draw a straight line between them with a ruler

STEP 3:

and place the baseplate on that line. Also, be sure the direction-of-travel arrow points in the direction you want to go.

4) With the map and baseplate held steady, turn the compass housing until the north arrow printed inside it lines up with the declination lines you drew previously.

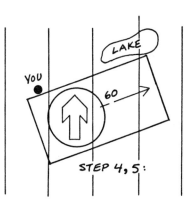

STEP 4, 5:

5) Read the bearing where the direction of travel arrow touches the compass housing. That's the bearing you'll follow when hiking.

6) Remove the compass from the map and follow the bearing (see page 218) you read in step 5. Remember that now is the first time you use the compass magnetic north arrow in this procedure.

Problem #2: You are hiking cross-country along a ridge and want to find out exactly where you are on the map.

Using Method #1:

1) Look around you until you see a prominent landmark like a hill.

2) Take a bearing (see page 218) to that hill with your compass. For the remaining steps keep the direction-of-travel arrow set at that bearing.

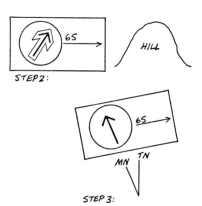

STEP 2:

STEP 3:

3) Lay the map flat on the ground. Then turn it until its magnetic north line lines up with the compass magnetic north needle.

4) Find the hill on the map. If you don't know which exact one it is, find several possible ones on the map.

5) With the compass magnetic north needle still coincidental with the map's magnetic north line, place the compass on your map so that its baseplate touches the hill you selected. Draw a line along the baseplate and extend that line to the edges of the map with a ruler. Repeat for each possible hill. Your position is somewhere on one of those lines.

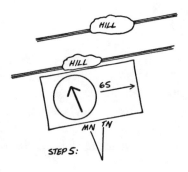

STEP 5:

6) Now select a second prominent landmark like a lake and repeat steps 2 to 5 to obtain a *cross-bearing*. Your exact position is where both lines intersect. If you used several possible landmarks in step 4, you'll have several intersection points. Use the map's topographical features like contour lines to determine which position is actually yours.

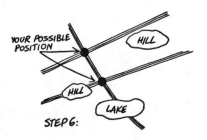

STEP 6:

Using Method #2:

1) Look around until you see a prominent landmark like a hill.

2) Take a bearing (see page 218) to that hill with your compass. For the remaining steps keep the direction of travel arrow set at that bearing.

3) Lay the map flat on the ground with no particular orientation.

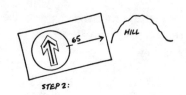

STEP 2:

4) Find the hill on the map. If you're not sure which one it is, find several possible ones on the map.

5) Place the compass on the map so that the north arrow printed inside the housing lines up with the declination lines you drew previously, so that the bearing you took in step 2 lines up with the direction of travel arrow, and so that the compass base-plate touches the hill you selected.

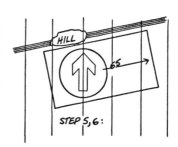

6) Draw a line along the base-plate and extend it to the edges of the map. Repeat for each possible hill. Your position is somewhere on that line.

7) Now select a second prominent landmark like a lake and repeat steps 2 to 5 to obtain a *cross-bearing*. Your exact position is where both lines intersect. If you used several possible landmarks in step 4, you'll have several intersection points. Use the map's topographical features, like contour lines, to determine which position is actually yours.

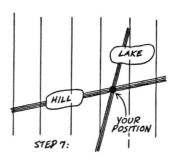

Problem #3: Use a map and compass to identify a mountain you see in the distance.

Using Method #1:

1) Lay your map flat on the ground. Then turn it until its magnetic north line lines up with the compass magnetic north needle.

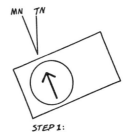

2) Mark your position on the map. Use the steps from Problem #2 to find your position if necessary.

3) Take a bearing (see page 218) to the mountain with your compass. For the remaining steps, keep the direction-of-travel arrow lined up with that bearing number.

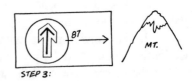

STEP 3:

4) Place the compass on the map so that any corner of the baseplate touches your position. Then pivot the entire compass around your position until the north arrow printed inside the housing lines up with the magnetic north needle.

5) The mountain lies along the line formed by the baseplate and in the direction indicated by the direction-of-travel arrow. It can be identified by reading the symbols like contour lines on the map.

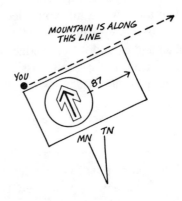

Using Method #2:

1) Lay the map flat on the ground with no particular orientation.

2) Mark your position on the map.

3) Take a bearing (see page 218) to the mountain with your compass.

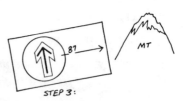

STEP 3:

4) Place the compass on the map so that the north arrow printed inside the housing lines up with the declination lines you drew previously, so that the bearing you took in step 2 lines up with the direction-of-travel arrow, and so that a long side of the baseplate touches your position.

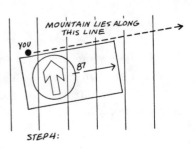

STEP 4:

5) The mountain lies along the line formed by the baseplate and in the direction indicated by the direction-of-travel arrow. It can be identified by reading the symbols like contour lines on the map.

Deliberate Error

When using your map and compass for cross-country hiking, use the concept of *deliberate error* to reach your destination. Look at Figure 1 in Example 17-10. If you parked your car along a dirt road and wanted to hike through thick woods to a fishing lake that lies 40° northeast of you, don't hike directly to it. If your bearing measurement is inaccurate by as little as 2–3°, you'll miss the lake and get lost. Instead, aim a little more to the north on a bearing of 38–40°, intersect the stream, and follow that to the lake. On your return trip, don't walk directly back to the car on a bearing of 220° (40° plus 180° for an opposite direction). You could be a few degrees off course and reach the road but not know if the car is to the left or right on it. Instead, follow a bearing a little more to the right. That way, when you reach the road, you'll know for sure that the car is down the road to your left.

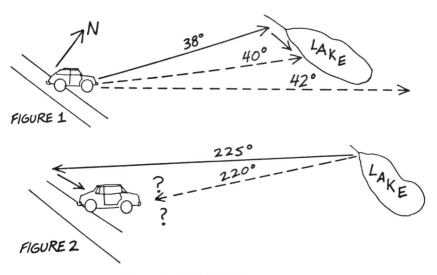

Example 17-10: Deliberate error.

Estimating Time and Direction

The following methods for measuring time and direction are relatively accurate most of the time in most places in the continental United States. All times mentioned in the next section are Standard, not Daylight Saving Time.

With a Bright Sun or Moon

1) *Direction*—Put a stick straight up in the ground and mark the tip of its shadow with a pebble. Wait at least 15 minutes, mark its shadow again with another pebble, and draw a straight line connecting the pebbles. That line points directly east and west. (With the sun or moon at your back, east will be on your right and west on your left.) This method is most accurate when the sun or moon are not very close to the horizon (Figure 1, Example 17-11).

2) *Direction and time*—Put a stick vertically in the ground and mark its shadow at 15-minute intervals throughout the middle part of the day. When the shadow is the shortest length, it's 12 o'clock and the sun or moon is directly south. A line connecting the ends of the shadows runs directly east and west (Figure 2, Example 17-11).

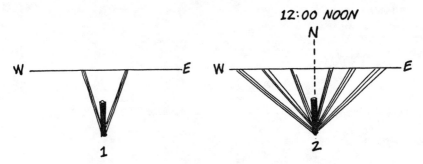

Example 17-11: Using sticks and shadows to determine direction and time.

3) *Time*—To estimate how soon the sun or moon will set, hold your hand straight out at either of them with your fingers curled around between it and the horizon. If the sun or moon is one finger away from the horizon, it will set in about 15 minutes. If it's four fingers away from it, it will set in about 1 hour (Example 17-12).

4) *Direction*—In March and September, the sun rises almost directly east and west. During April to August, it rises in the northeast and

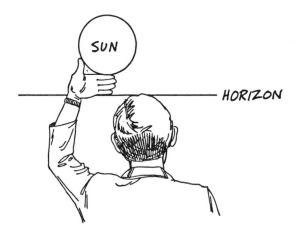

Example 17-12: Estimating how soon the sun or moon will set.

sets in the northwest. Between October and February, it rises in the southeast and sets in the southwest.

5) The moon rises about an hour later each day.

With a Compass

1) *Time*—When the sun is directly east, it's 6 a.m. and when it's directly west, it's 6 p.m. You'll have to estimate the time during the winter months when the sun rises after 6 a.m. and sets before 6 p.m. When the sun is directly south, it's noon.

2) *Time*—Place a compass on level ground in the sun, orient it so it's magnetic needle lines up with the needle inside its housing, and adjust for the declination. Then hold a pencil or thin stick straight up on its center point. When the stick's shadow lines up with north on the compass, it's noon, when it lines up with east it's 6 p.m., and when it lines up with west, it's 6 a.m. Make a small sun dial diagram to find the other times (Example 17-13).

With a Wristwatch (with Correct Local Standard Time)

1) *Directions*—At 6 a.m. the sun is directly east and at 6 p.m. it's directly west.

2) *Directions*—Lay your watch down flat on the ground with the hour hand pointing at the sun. South lies halfway between the hour hand and 12 o'clock on the watch (Example 17-14).

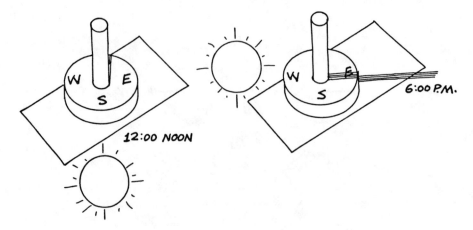

Example 17-13: Telling time with a compass.

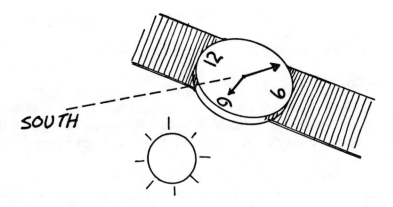

Example 17-14: Finding directions with a wristwatch.

Finding the North Star

Contrary to popular belief, the North Star is not the brightest and most obvious star in the sky. In fact, it's difficult to find unless you know where to look for it. Use an astronomy book to study the stars and learn how to identify the North Star because in a survival situation, finding the North Star will be one of your least reliable methods of finding the directions if you don't know where to begin to look for it. If you do need to use the North Star for finding directions and don't know how to find it, follow these suggestions:

1) Use a compass to point in the general direction of north, and then find the North Star by using the "bright star method" described below. Of course, if you have a compass, you only really need to find the North Star when you must walk on an exact compass bearing and have forgotten the magnetic declination for that area. The North Star always points within 1° of true north, while a compass can point as much as 30° to the east or west of true north in the continental United States, depending on your location.

2) Since the moon rises in the east, sets in the west, and follows the same path as the sun across the southern part of the sky, use it to find the general direction of north with one of the previously described methods. Then look for the bright guide stars described in Example 17-15 to help locate the North Star.

3) Look around the whole sky (or just in the northern part if you know where that is) until you find the pattern of stars drawn in Example 17-15. They're very bright and should stand out even in mildly cloudy weather. Depending on the time of night and year, they could be in

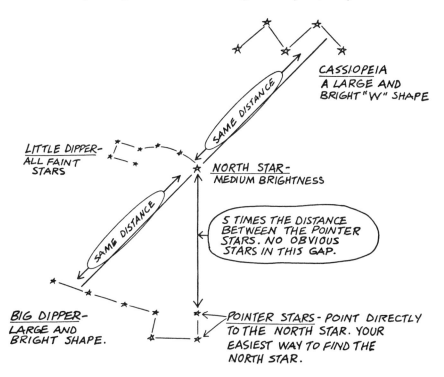

Example 17-15: Finding the North Star.

Example 17-16: Lining up two sticks with the North Star to record north for later use.

different positions than that shown in the drawing, but the positions of the stars in relation to each other will always be the same.

After you've found the North Star, line up two objects like sticks with it to record true north for later use (Example 17-16).

18

Ropes and Knots

There are several major kinds of rope. *Manila* rope splinters, rots when wet, is fairly heavy, and has a low breaking strength. *Cotton* rots and breaks easily. *Polypropylene* is a plastic rope useful for canoe camping because it floats. *Binder's twine* is inexpensive cord that disintegrates quickly. *Nylon* is strong, light, rot- and mildew-proof, and durable but stretches a great deal when supporting heavy weights. Nylon is the most popular kind of rope used for backpacking, although some people use *perlon* or *dacron* which are similar to nylon but don't stretch nearly as much.

A rope's *test strength* is the amount of weight it can hold without breaking under ideal conditions. A rope's *safe working load* is one-quarter to one-fifth its test strength. A rope rated at 600 pounds test will safely support a weight of 150 pounds in a camping setting if used properly, although jerking the rope doubles the strain on it, and a knot in it decreases its safe working load by well over 50%.

Care

The rope you need for backpacking requires very little care. Don't walk or stand on your rope, and keep it as clean as possible. Coil a large amount of rope between your palm and elbow and coil a small piece around your palm for storage and packing (Example 18-1).

You need to *whip* the ends of a rope to prevent it from unraveling. For nylon rope, simply hold a match under the end of it until it melts. To

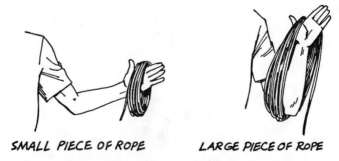

SMALL PIECE OF ROPE   LARGE PIECE OF ROPE

Example 18-1: Coiling a rope for packing.

keep the rope ends smooth and not lumpy, scrape the lump of liquid nylon off with a stick before it hardens. Whip other kinds of rope by wrapping their ends with tape or by the method described in Example 18-2.

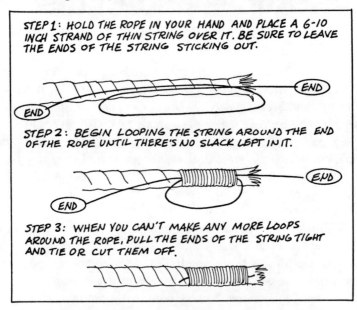

STEP 1: HOLD THE ROPE IN YOUR HAND AND PLACE A 6-10 INCH STRAND OF THIN STRING OVER IT. BE SURE TO LEAVE THE ENDS OF THE STRING STICKING OUT.

END

END

STEP 2: BEGIN LOOPING THE STRING AROUND THE END OF THE ROPE UNTIL THERE'S NO SLACK LEFT IN IT.

END

END

STEP 3: WHEN YOU CAN'T MAKE ANY MORE LOOPS AROUND THE ROPE, PULL THE ENDS OF THE STRING TIGHT AND TIE OR CUT THEM OFF.

Example 18-2: Whipping the ends of a rope.

Knots

Practice tying the knots on the following pages until you know them well because they're the knots you'll need for almost all backpacking trips. You know you've mastered these knots when you can tie them behind your back with your eyes closed.

To untie a stubborn knot, wedge your pocket knife awl or screwdriver blade between the strands and pry (not cut) them apart. You'll loosen almost any knot by carefully working at it for a while this way (Example 18-3).

Example 18-3: Using your pocket knife awl to loosen a stubborn knot.

Safety: While this is all you need to know about ropes and knots for general backpacking, there obviously is a lot more to know. Before you use any rope to rescue anyone, to support a person's weight for climbing, or for an extended backpacking trip in rugged terrain, study a technical book written entirely on the subject.

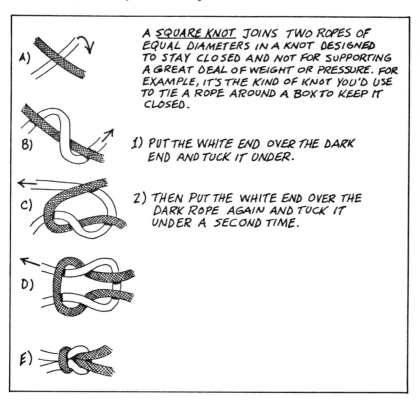

A SQUARE KNOT JOINS TWO ROPES OF EQUAL DIAMETERS IN A KNOT DESIGNED TO STAY CLOSED AND NOT FOR SUPPORTING A GREAT DEAL OF WEIGHT OR PRESSURE. FOR EXAMPLE, IT'S THE KIND OF KNOT YOU'D USE TO TIE A ROPE AROUND A BOX TO KEEP IT CLOSED.

1) PUT THE WHITE END OVER THE DARK END AND TUCK IT UNDER.

2) THEN PUT THE WHITE END OVER THE DARK ROPE AGAIN AND TUCK IT UNDER A SECOND TIME.

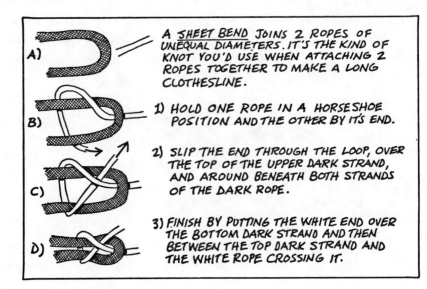

A _SHEET BEND_ JOINS 2 ROPES OF UNEQUAL DIAMETERS. IT'S THE KIND OF KNOT YOU'D USE WHEN ATTACHING 2 ROPES TOGETHER TO MAKE A LONG CLOTHESLINE.

1) HOLD ONE ROPE IN A HORSESHOE POSITION AND THE OTHER BY IT'S END.

2) SLIP THE END THROUGH THE LOOP, OVER THE TOP OF THE UPPER DARK STRAND, AND AROUND BENEATH BOTH STRANDS OF THE DARK ROPE.

3) FINISH BY PUTTING THE WHITE END OVER THE BOTTOM DARK STRAND AND THEN BETWEEN THE TOP DARK STRAND AND THE WHITE ROPE CROSSING IT.

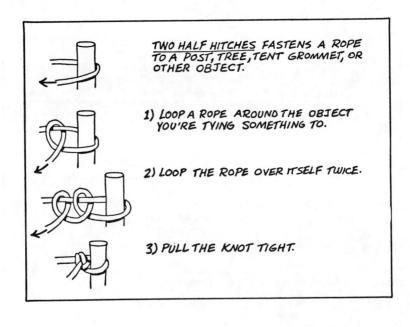

TWO HALF _HITCHES_ FASTENS A ROPE TO A POST, TREE, TENT GROMMET, OR OTHER OBJECT.

1) LOOP A ROPE AROUND THE OBJECT YOU'RE TYING SOMETHING TO.

2) LOOP THE ROPE OVER ITSELF TWICE.

3) PULL THE KNOT TIGHT.

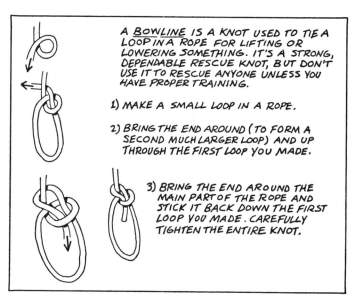

A <u>BOWLINE</u> IS A KNOT USED TO TIE A LOOP IN A ROPE FOR LIFTING OR LOWERING SOMETHING. IT'S A STRONG, DEPENDABLE RESCUE KNOT, BUT DON'T USE IT TO RESCUE ANYONE UNLESS YOU HAVE PROPER TRAINING.

1) MAKE A SMALL LOOP IN A ROPE.

2) BRING THE END AROUND (TO FORM A SECOND MUCH LARGER LOOP) AND UP THROUGH THE FIRST LOOP YOU MADE.

3) BRING THE END AROUND THE MAIN PART OF THE ROPE AND STICK IT BACK DOWN THE FIRST LOOP YOU MADE. CAREFULLY TIGHTEN THE ENTIRE KNOT.

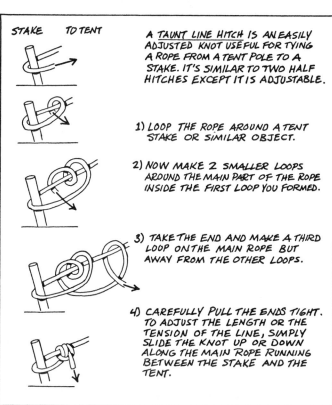

STAKE   TO TENT

A <u>TAUNT LINE HITCH</u> IS AN EASILY ADJUSTED KNOT USEFUL FOR TYING A ROPE FROM A TENT POLE TO A STAKE. IT'S SIMILAR TO TWO HALF HITCHES EXCEPT IT IS ADJUSTABLE.

1) LOOP THE ROPE AROUND A TENT STAKE OR SIMILAR OBJECT.

2) NOW MAKE 2 SMALLER LOOPS AROUND THE MAIN PART OF THE ROPE INSIDE THE FIRST LOOP YOU FORMED.

3) TAKE THE END AND MAKE A THIRD LOOP ON THE MAIN ROPE BUT AWAY FROM THE OTHER LOOPS.

4) CAREFULLY PULL THE ENDS TIGHT. TO ADJUST THE LENGTH OR THE TENSION OF THE LINE, SIMPLY SLIDE THE KNOT UP OR DOWN ALONG THE MAIN ROPE RUNNING BETWEEN THE STAKE AND THE TENT.

19

Predicting
the Weather

While the following indicators of fair and foul weather
are accurate only about 75% of the time, they'll nonetheless give you an
idea of possible dangerous weather conditions so you can prepare for or
avoid them. The more of these signs you observe, the more reliable the
prediction will be.

Signs of Fair Weather

. . . . . many fresh spider webs.
. . . . . smoke rises steadily upwards.
. . . . . a red sky in the evening.
. . . . . a heavy morning dew indicates no rain before the afternoon.
. . . . . a heavy evening dew indicates no rain before dawn.
. . . . . a clear, bright moon.
. . . . . a rainbow in the evening or a rainbow to leeward.
. . . . . cold, clear weather after a storm lasts longer than mild
weather following a storm.
. . . . . morning fog will burn away before noon.

. . . . . cumulus clouds (which look like puffy cotton balls) . . .
—a regular pattern of clouds as well as regularly shaped
  clouds,
—scattered clouds with a lot of blue sky between them,
—the higher the clouds, the fairer the weather,
—remain a constant shape as they drift across the sky.
. . . . . birds flying late in the evening.
. . . . . a clear evening sky indicates fair weather and a very cold
  night.

Signs of Approaching Uncertain or Stormy Weather

. . . . . smoke rises a short way and then settles back down.
. . . . . a red sky in the morning.
. . . . . a ring around the moon (especially if it gets smaller). However, a very large ring around the moon is caused by ice
  crystals in the upper atmosphere and doesn't necessarily
  signify stormy weather.
. . . . . leaves show their undersides.
. . . . . flower blossoms close.
. . . . . sounds are clearer and travel farther as a storm approaches.
. . . . . the wind increases.
. . . . . no morning dew indicates rain before evening.
. . . . . no evening dew indicates rain before dawn.
. . . . . birds and insects fly lower to the ground. Rows of birds
  perched on trees or powerlines indicate an approaching
  storm.
. . . . . use your senses. Some people can smell rain coming.
. . . . . a rainbow in the morning or a rainbow to windward.
. . . . . heat lightning, which is lightning that's too far away to hear
  its thunder, indicates stormy weather within 10–30 miles. It
  may be heading your way or just passing around you.
. . . . . sharp horns on the moon and very distinct stars are signs of
  a sky unusually free of atmospheric moisture and dust. A
  very clear sky indicates windy weather and a change in
  storm fronts that could (but often doesn't) bring rain.
. . . . . expect a long storm when the rain comes before the wind.
  Expect a shorter storm when the wind comes before the
  rain.

..... fog that settles before darkness will linger until after the
dawn.

..... a storm observed in the west or the northwest will probably
hit you, but storms observed in other directions will probably miss you.

..... clouds . . .

—cumulus clouds build up vertically, their lower parts
darken, or their tops flatten out (Example 19-1),

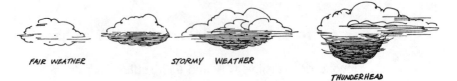

FAIR WEATHER   STORMY WEATHER

THUNDERHEAD

Example 19-1: Cumulus clouds.

—the higher the thunderheads build up, the greater the rain-
fall expected,

—clouds move in different levels and layers, which indicates an unstable atmosphere,

—increasing amounts of cirrus clouds (which look like
feathers or paintbrush strokes high in the sky) indicate
a storm will reach you in about 15 hours,

—a general gloomy gray or unlayered mass of gray clouds
approaching indicates a long mild or moderate rainy
spell,

—a mackerel sky indicates precipitation within 15 hours
(Example 19-2).

Measuring Air Temperature

Measure the air temperature by counting the number of cricket
chirps in 14 seconds and then add 40 to that number.

Example 19-2: A mackerel sky.

IV
DANGER

20

Safety and First Aid

First aid is the temporary care you give someone when medical help is not available. In other words, first aid is helping a person as best you can when a doctor is not present; it's not playing doctor yourself. Always carry a first aid kit with you outdoors and know how to use it, but remember that carrying a kit is no substitute for the skills you'll learn in a Red Cross First Aid Course, which are necessarily very basic in nature themselves. First aid is a very complicated subject best learned from hours of reading and many more hours of practice and is impossible to learn from one chapter in a book. Thus, the information presented here stresses ways you can prevent common hiking problems and deal with specific hiking emergencies, rather than presents an exhaustive list of do's and dont's on how to treat every possible calamity that could affect you on the trail.

Prevention is the key that locks the door of danger. Don't get caught outdoors without it. Tell someone where you are going and when you expect to return. Carry an adequate amount of clothing, food, and water, and never eat all your food or drink all your water. Carry dependable shelter for the worst weather you could possibly encounter. Know the specific hazards, like dehydration and poisonous snakes, for the area you'll hike in. Know your body's physical limits and never exceed them, and always save some energy and strength for any emergencies that could arise.

Be extra careful *when camping on mountaintops* and exposed ridges. Check the five-day weather forecast or the weather signs on pages 240 and 241 to be sure there's no chance of a storm sweeping in at night, and in case an unexpected storm does blow in, be prepared to retreat to lower elevations immediately. Dress with warm outer garments designed to shed wind, and be especially prepared for hypothermia.

Be aware of the carbon monoxide poisoning and fire danger *when cooking inside your tent*. Allow plenty of ventilation and be careful not to knock the stove over. Even just a pot of spilled water could threaten your life if it soaks your sleeping bag in cold weather. Since carbon monoxide is an odorless, tasteless, colorless gas, its presence is often not known until too late. If you begin to get a headache, or feel giddy, sick to your stomach, or dizzy, turn off the stove and get fresh air immediately.

Avoid *dysentery* by thoroughly washing and sterilizing your cooking gear. Avoid digestive system upsets by completely removing soap residues when washing dishes. Always purify all drinking water you're not absolutely sure is safe.

When exploring unknown caves, carry at least three sources of light which could include flashlights, matches, candles, and cigarette lighters. Carry rope or toilet paper to mark your way, never go in alone, keep your group together so no one gets lost, look backwards often to memorize the way out, and never go in beyond side passageways that could confuse you when you return. Wear a hat to protect your head and long pants, long-sleeved shirts, and boots to protect the rest of your body. Cave temperatures are usually a constant 55° year-round, so wear appropriately warm clothes. (As an aside, cave formations are rare and fragile. Leave all formations alone. The oil from one fingerprint can destroy some cave formations.)

Never explore around or in old *mine shafts*, which are unstable and could collapse at any time.

Storms

Because light travels faster than sound, you can estimate how far away a *lightning storm* is by counting the time between the flash of lightning and the sound of the thunder. Every 5 seconds difference in time equals 1 mile in distance. For example, if you hear thunder 12 seconds after you see a lightning flash, that lightning struck aboout 2 2/5 miles away. Use this method to determine if the storm is moving closer to you, moving away from you, or just passing along your side.

If a lightning storm is approaching, avoid exposed hills, cliffs, isolated groves of trees, bodies of water, fencelines, powerlines, shallow caves, and overhanging rock ledges. Generally, the safest places are low and away from tall objects like trees or mountain ridges. Dense woods, large groves of young trees, or ravines offer the best protection in flat country. If you're caught in the open, move away from all metal objects and crouch low on your feet on any available insulation like a log, your sleeping bag, or your foam pad with as little of your body touching the ground as possible.

When a *sand or dust storm* overtakes you, cover your mouth and nose with a wet cloth and breathe slowly and deliberately. Seek shelter inside your tent, in the lee side of a boulder, or just with your back to the wind.

Foot Care

Toughen your feet before going on a demanding backpacking trip by walking barefoot on pavement, golf courses, or lawns. Strengthen your ankles by standing and walking on their edges. Put rubbing alcohol on your feet in the morning and evening during the week before a hike to toughen your skin. Trim your toenails short and straight across. Wear your hiking boots during work or around your house at night a week before a trip to get your feet accustomed to wearing them. Above all, never begin a backpacking trip with brand new boots.

Wash your feet every day whenever possible on a hike, or at least air out your feet and massage them several times a day if you can't wash them. Use foot powder to prevent athlete's foot and reduce friction that causes blisters. Carry 1–2 ounces of rubbing alcohol and apply it to your feet to cool them when hiking in hot, dry terrain.

Most *blisters* form when hiking downhill, so tighten your boots at the top of a long hill to keep your feet from sliding inside your boots and causing blisters. Be sure your socks fit snugly but without restricting your toes and without wrinkles, holes, and worn spots. If there's a place on your feet where you frequently get blisters, cover it with an adhesive bandage, moleskin pad, or adhesive tape. Readjust your socks and remove any stones in your boots immediately if they cause problems.

If you find a small blister on your foot, cover it with an adhesive bandage or gauze pad but not with moleskin, which sticks to blisters and is very hard to remove. If a large blister develops, pop it if you have much more walking to do that day. Sterilize a needle in a match flame, prick the blister from the side, drain it completely, dry it with a gauze pad, and finally cover it with an adhesive bandage or gauze pad. At the end of the day, remove the bandage, air-dry the blister in the sun to help sterilize it, and cover it with a fresh bandage.

After blisters, a turned ankle is probably the most common problem that could affect you on a backpacking trip. The severity of an ankle injury can be roughly estimated from the amount of swelling and pain present and the difficulty you have walking on it. However, this is a general guideline only. While serious ankle injuries are rare, they occasionally occur without severe swelling, pain, and limb impairment. The standard first aid procedure for treating ankle injuries is immobilizing *all* injuries, but this is often impractical outdoors, especially since most twisted ankle injuries are very minor. To be on the safe side, though, never walk on an injured ankle unless you are sure the injury is extremely minor. For a mildly twisted ankle, apply ice or cold water for a while and then cautiously proceed with your hike. First aid for severely twisted ankles includes applying ice or cold water as soon as possible and for as long as 24 hours, mild compression of the joint to reduce swelling, elevation of the ankle above the hips, aspirin to relieve any pain, and rest (that means don't hike on it). An elastic bandage is ideal for applying compression but always begin wrapping it from the toes upward and never wrap it so tight that circulation is impaired. Blue not pink toes indicate poor circulation. If you have to hike on a severely injured ankle, immobilize it and support it with a bandana or shirt securely wrapped around it and around and under your hiking boot. Consult a doctor when possible for serious injuries.

Immersion foot occurs when your feet are constantly wet from water or sweat. Your feet look white and shriveled and feel cool and damp at first but eventually become numb and lose their feeling. Prevent it by wearing proper footgear in wet climates, and by keeping your feet dry by changing socks, airing out your feet, and massaging them for 5-10 minutes several times a day.

Altitude Dangers

Altitude sickness occurs when people from low altitudes make a rapid ascent to a high elevation (usually over 6,000 feet). Its symptoms include nausea, insomnia, headache, dizziness, shortness of breath, and coughing, and is more of an uncomfortable problem than a life-threatening danger. Treat it by reducing your physical activity and resting more often at the higher altitude or by retreating to a lower elevation.

High-altitude pulmonary edema is a dangerous reaction to high altitudes caused by excessive fluid collecting in the lungs. Its symptoms include a bluish skin color, coughed-up blood, gurgling sounds in the lungs, shortness of breath, rapid pulse, dry and persistent cough, severe headache, hallucinations, and general body weakness. Treat it by rapidly descending to lower elevations and then seeking medical help immediately. This reaction is rare below 10,000 feet.

Water Dangers

Swimming

Swimming is a real treat on a summer backpacking trip, but if you do swim, swim safely. Never dive or jump into a river or lake if you can't see the bottom. Wear sneakers or boots to protect your feet from sharp rocks, broken glass, and fishhooks. If you swim barefoot, be sure you can see where you put your feet, and be extra careful if you swim alone. Stay near shore where you can crawl out if you get cramps, are bitten by poisonous snakes (in the southeastern part of America), or get numb in cold water. If you're with a group of hikers, have someone act as a lookout to watch for anyone in trouble, or have everyone pair up with another person as buddies for safety. Children often exceed their physical limits when swimming. Be especially aware of them in and near water.

If you need to rescue someone having trouble in the water, follow these rules in this order:

1) *Reach* something, like a shirt, stick, belt, or log out to the person in trouble.

2) *Throw* something out to him, like a rope or a floating log.

3) *Row* a boat or a log out to him.

4) *Go* in the water to rescue him yourself *only if there is no other way to save him, and only if you know how to rescue a drowning person.*

Don't be a hero. There's a whole Red Cross Lifesaving Course that teaches water rescue techniques. Take the course before going for a medal.

Temperature Dangers

The human body is a furnace with two main parts. Its *core* is its head and chest area containing vital organs like the brain, lungs, heart, and liver. For the body to function properly, the core temperature must be maintained at a constant level within a degree or two of 98°F. The *shell* is the outer layer of skin and the extremities. Those noncritical areas of the body can become very cold or warm without affecting the body's overall operation or survival.

When cold, your body automatically begins to conserve its heat in the following ways:

1) Your hands, toes, and outer layer of skin get cold because the blood supply that normally goes there concentrates in the core instead.

2) Your body shivers, which is its way of exercising to produce more heat.

3) The body's metabolic rate increases and converts stored body fat to energy.

When warm, your body gets rid of excess heat in these ways:

1) Sweating acts like an air conditioning system that cools it through evaporation.

2) Blood vessels near the skin dilate to carry excess warmth from the core to the outside environment.

Hypothermia

Hypothermia is the general cooling of the body's inner core temperature. It occurs when the heat lost from your body is greater than the heat produced by it for long, gradual periods or for short, rapid periods of time. Hypothermia is fatal to people falling into cold arctic water in as little as 10 minutes, but could take up to two to three days for someone stranded with no shelter in 50° weather. It can occur at any altitude from sea level to mountaintops, in any temperature (but usually between 30–50°F) and during any time of the year. It's the major cause of the death of people engaged in recreation outdoors.

Hypothermia is caused by numerous compounding and multiplying factors including inadequate protection from the wind, rain, and cold, fatigue from too much exercise, and inadequate food and water. Technically, it occurs in three progressively worsening stages.

In the first stage, your body core temperature ranges between 95–98°F. You shiver, you feel chilled, and your fingers and toes get cold. Your body is trying to generate more heat and concentrate all available heat in its core. This stage is common to many people including campers, hunters, and spectators watching football games on cold autumn evenings. This is a warning stage. While not dangerous in itself, it could lead to the next stage if not treated within a reasonable amount of time.

In the second stage, your core temperature drops to somewhere between 90–95°F. Two major symptoms occur in this stage that you should never ignore. The first is the loss of the use of your hands. When camping this means that you become virtually helpless since you can't put on warmer clothes or build a fire. The second major symptom is that your body's uncontrollable shivering usually stops, which is a signal that it has used up all its available fuel for heat production. Now the core temperature will plummet unless you get proper shelter and warmth immediately. Other symptoms of this stage include slurred speech, amnesia, poor judgement, very numb fingers and toes, and poor overall coordination.

As your body's core temperature continues to drop below 90°F in the third and final stage, you become unconscious, your pulse and breathing rates lower, and you become totally dependent on other people to save you. When your core temperature drops to about 85°F, you'll die.

You can prevent hypothermia by always carrying a dependable shelter to get out of the wind and rain, plenty of high energy food to maintain an active metabolic rate, and extra dry clothes. Dress for adequate protection from the wind, rain, and cold by wearing synthetics or wool instead of cotton, a hat that reduces heat radiated from your head and neck areas, and windproof garments over your inner insulation layers.

Treat a hypothermia victim by preventing any further heat loss and then gradually adding heat to warm him up. Keep the victim out of the wind, rain, and snow, replace his wet clothes with dry ones, and give him warm drinks and candy or high energy foods if he's conscious. Place him inside a sleeping bag with another person, or have him sit by a fire or huddle in the middle of everyone else in the group. Never put a person suffering from hypothermia in a cold sleeping bag alone, since his body won't generate enough heat to warm it up.

Wind Chill

A *wind chill chart* (Example 20-1) tells you the temperature your bare skin feels at different thermometer readings and wind speeds. To use the chart, estimate the wind speed (by using Example 20-2) and the air temperature. The *wind chill factor*, the temperature it feels like at a specific air temperature and wind speed, is where the lines intersect. For example, if a thermometer records a temperature of 40° and a 10-mile-an-hour wind is blowing, your exposed skin feels a temperature of 28°. You could be in serious trouble if you were hiking under those conditions and didn't carry clothes that could keep you warm below freezing.

	temperature						
wind speed	*50*	*40*	*30*	*20*	*10*	*0*	*-10*
0	50	40	30	20	10	0	-10
5	48	37	27	16	6	-5	-15
10	40	28	16	4	-9	-21	-33
15	36	22	9	-5	-18	-32	-45
20	32	18	4	-10	-25	-39	-53
25	30	16	0	-15	-29	-44	-59
30	28	13	-2	-18	-33	-48	-63
35	27	11	-4	-20	-35	-51	-69

Example 20-1: The wind chill chart.

wind speed	**observable signs**
5	light breeze, leaves rustle
10	small branches move
15	heavy ripples on water, large branches move
20	small trees move
25	large waves on water, larger trees sway

Example 20-2: Estimating wind speed.

Water Chill

Since water conducts heat much faster than dry air, it's imperative to stay as dry as possible during colder weather. Water chill, especially when combined with a wind chill, can kill you in an incredibly short time.

Heat

In warm weather, you often must take extra measures to help your body get rid of unneeded heat. Wear light-colored clothes which reflect the sun's warmth away from you and loose-fitting ones that let perspiration escape. When water is plentiful, soak your clothing in it to increase the amount of heat conducted and evaporated from your body. Even a wet bandana tied loosely around your neck cools you considerably in hot weather. Drink plenty of water, reduce your physical activity, seek any available shade, don't eat much solid food to lower your metabolic rate, and don't travel during the hottest part of the day. If you overexert yourself and don't take measures to maintain a cool body temperature, your body's core temperature will gradually rise and you'll suffer from heat exhaustion, heat stroke, or dehydration.

You'll *dehydrate* if you don't drink enough water. Dehydration is a very insidious problem, because it can develop in a few hours on a hot day or over a period of a few days in cooler weather. The most obvious signs of it are infrequent, dark-yellow urine and a dry mouth. Other symptoms include irritability, nausea, dizziness, poor judgement, and poor overall coordination.

If you sweat profusely in hot weather, you could develop a *salt deficiency* that causes cramps and nausea. Prevent it by eating a sufficient amount of salt with plenty of water.

Heat exhaustion is your body's response to heat caused by insufficient water and excessive heat. Too much blood concentrates at the skin's surface to get rid of excessive internal heat and not enough remains in the core for the body to function properly. Symptoms of heat exhaustion include fatigue, weakness, a faint feeling, and white, cool, or clammy skin. Treat it by laying down for about an hour, applying cold wet cloths to your skin, and getting out of the sun and heat. Relax and take it easy for a few days after recovering. Further medical treatment is seldom necessary.

Heat stroke is a much more serious condition that signals the body's inability to cool itself any longer. It's symptoms include a very high temperature (often greater than 106°), hot, red, and dry skin, a very rapid and strong pulse, and unconsciousness. Treat it by immediately cooling the victim's skin with cold water baths, fanning it to increase evaporation and convection heat losses, getting him out of the sun, and taking him to a doctor as soon as possible.

Animals

The dangers from wild animals are greatly exaggerated and virtually nonexistent outdoors. With rare and excessively publicized exceptions, no animal will attack unless provoked, cornered, separated from its young, or approached during the mating season.

Skunks are mellow creatures that will sneak up in the shadows and darker places around your campsite at night. Observe them from a distance or approach them for a closer look if you want to. As long as you don't startle them, they'll almost always warn you they're getting angry before spraying. First they'll stare at you and stamp their front feet on the ground. Then, if they get more upset, they'll turn around and raise their tail in the air. Finally, if they still think you're threatening them, they'll spray. If sprayed, repeated washings in tomato juice is the best way to remove the smell from your skin and clothing.

Porcupines will occasionally wander through your campsite and chew on anything containing salt, including sweaty walking sticks, gloves, and clothes. You can approach very close to them because the only way you can get the quills imbedded in you is to actually touch one. They don't attack people and don't shoot quills in the air.

Raccoons will appear at night looking for food. Sometimes they'll spend all evening trying to reach your food supply hanging in a tree (see page 260). If you hear a loud, constant chattering sound around your food cache, it's probably a group of raccoons. They're mean fighters when cornered, but avoid people if left alone.

Coyotes are curious, harmless animals not overly scared of people. They'll approach very close to you and will walk around your campsite when you're asleep, but won't bother you if you leave them alone. Their melancholy song is one of the pleasures of the outdoors.

Mountain lions occasionally follow hikers out of curiosity but remain at a safe distance and only attack if cornered. The last thing you should worry about when camping is being attacked by a mountain lion though, because human sightings of them and contact with them are extremely rare.

Although *elk* and *moose* ignore people if left alone, avoid them in the fall mating season and especially if accompanied by their young. If attacked, climb a sturdy tree or find a thicket of tangled trees too dense for them to enter.

Squirrels, chipmunks, gophers, rabbits, mice, rats, and all kinds of *birds* will sneak into your campsite looking for food and are the animals you'll have to deal with most often. Avoid the temptation to feed or pet them, because they could get too close and bite your hand by mistake. From a distance of several feet, they're harmless.

Other animals like *wolves, bobcats, foxes, weasels, deer* and *otters* are harmless and avoid contact with people.

In certain areas *domestic or wild dogs* will threaten you much more than wild animals. When you meet an aggressive dog, find a club for defense, throw stones at it, or climb a tree and wait until it loses interest in you. Just the act of bending over to pick up a stone scares off some dogs. If you're going to be hiking through a lot of farmlands or small towns, consider carrying a 2-ounce can of *chemical self-defense spray* available in stores for safety.

Cattle are common throughout much of the west, but pose no threat to hikers at all. They have learned that humans mean trouble to them and will move out of your way as soon as they notice you. A cowboy "yee-haa!!" yell is often enough to send a herd of cattle in retreat. If a bull doesn't move out of your way when it sees or hears you, simply walk around it with about 10–30 yards of space between you and it. It will leave you alone.

The *human animal* is often a very dangerous, unpredictable creature. Always wear bright clothing during hunting season, never leave your equipment unattended to prevent theft, and be wary of people who seem somewhat "strange." Fortunately, though, the farther you are from developed roads and campsites, the more you can trust the people you meet.

Bears

Grizzly bears and black bears are the two kinds of bears you could encounter in the continental United States. *Black bears* are relatively small in size, have a mild temperament, and live in most forested areas throughout North America. Black bears usually aren't aggressive and retreat when approached, because they simply want to be left alone. Usually they'll stand their ground or charge only if you get between them and their cubs or if you come upon them suddenly and without warning. Their attacks on hikers are usually limited to one swipe or bite, rather than a physical mauling typical of grizzly bear attacks.

If you suddenly meet a black bear outdoors, talk softly and gently to it to calm yourself and inform it of your presence. Then, if it doesn't retreat as soon as it notices you, slowly back away from it to let it know you don't want to challenge it. If you're hiking on a trail, make a wide detour around it or retreat the way you came and wait a while before proceeding again. If you hear a black bear sniffing around your tent or sleeping bag at night, talk softly to let it know you're there. Don't make sudden movements or loud noises which could startle it. In almost every case, black bears will quickly retreat as soon as they notice you. If a bear near your campsite doesn't retreat immediately after he notices you, throw several stones near (not at) it, scream at it, or bang pots and pans together to make a loud noise to scare it away.

A black bear that dips its head low along its legs is submissive and wants to avoid a confrontation, but one that growls, has its ears laid back, or advances towards you is angry and could attack. Many experienced outdoorsmen recommend dropping your backpack and running from an aggressive black bear, because in almost every case, the bear simply wants you away from their territory, their young, or their food (what they consider their food could include the food in your pack or at your campsite!). They'll seldom chase you when they see you running away. Other people recommend dealing with aggressive black bears by dropping your pack and slowly backing away from it while constantly talking in a low voice to calm it.

If a black bear catches you, don't scream or fight with it because that'll make it more angry. Lay on the ground, cover your head with your arms, roll into a ball, and act dead. Generally, black bears are passive creatures that avoid confrontations and rarely attack people without provocation. They simply want to be left alone.

Grizzly bears have a small head, a large body, a hump on their shoulders, a black, brown, or gold-brown coat, and a "dishpan," pressed-in face. They live in Alaska, Canada, and parts of Montana, Wyoming, Idaho, and northeast Washington. They're much more aggressive and unpredictable than black bears. While black bears charge to scare you away from their cubs or food, grizzly bears charge to kill you.

Because they're much larger and more dangerous than black bears, you must take extra precautions in grizzly bear country. Be alert at all times when hiking. Tie a bell to your pack, sing, or tap dishes together as you walk to announce your presence and give any bears in the area time to retreat. Never hike alone in grizzly country, because they confront solo hikers more than groups of people. In Yellowstone National Park, for example, there are signs at many trailheads warning people to hike in groups of at least four people. When attacked or confronted by a grizzly, drop your pack and climb the nearest tree, because grizzlies can't climb them. Never run away from a grizzly bear. For added safety, camp in a tent (not under a tarp or in the open) at least 200 yards upwind from your cooking area and food cache, and set up the tent near a tree you can easily climb in an emergency. Never travel at night in grizzly country.

Animals and Food

Generally, there are two kinds of animals. Truly wild animals live in areas far from human contact; they fear people and avoid them. The other kind, the kind typically found in national parks and near popular campsites, have lost their natural fear of man. Because of ignorant people who feed them and leave garbage at campsites, they've learned that contact with people poses no threat to them and in fact means food. Those are the kinds of animals that'll cause you all kinds of problems outdoors because they aren't scared of you.

While wild animals can survive and remain healthy on the natural food found in their habitat, many of them have discovered that it's easier digging for garbage, waiting for handouts, or raiding campsites than it is hunting and gathering wild food. In addition, the supermarket food and garbage they find actually becomes an addiction to them much like candy is for small children. They'll go to extreme measures to get it to satisfy their intense craving for it. For example, some bears in the Smoky Mountains actually have learned to approach hikers, scare them enough to drop their backpacks, and then drag their packs off into the woods to eat the food in them at their leisure! Anytime a bear takes a backpack, it considers it his. *Retake it at your own risk.*

When camping anywhere (but especially in bear country), keep a very clean cooking area and campsite so you don't attract unwanted visitors. Wash your dishes thoroughly, and don't spill or discard any food, dirty dish water containing food, toothpaste, soap, suntan lotion, or other "smellables" near your campsite. Try to burn leftovers in your campfire, but if you can't then discard them several hundred yards away. Never wipe your hands on your clothing when cooking, because they can pick up the food smell and attract wildlife. In fact, be very careful that your pack, tent, sleeping bag, and all other gear doesn't pick up the smell of spilled food around the campsite or from the fire's smoke while cooking.

After you're done eating dinner and before darkness sets in, separate all your smellables, including food, litter, clothes stained with food, suntan lotion, toothpaste, soap, medicines, and water bottles stained with flavored drink mixes. Pack all those things in your backpack, large plastic bag, nylon stuff sack, or other strong waterproof sack, and then carry it at least 100 feet from your campsite in black bear country, and at least 200 feet away in grizzly bear country. Find a tree with a sturdy branch that sticks out at right angles at least 15–20 feet above the ground. Tie a rock to your rope, throw it over the branch, tie the other end to your food sack, and hoist it up in the air. Keep it far from the trunk, branches, and ground so no bear can reach it. Secure the end by tying it around a convenient tree branch (Example 20-3). In areas not inhabited by bears you can safely cache your food as little as 5 feet from your campsite and can safely hang it just a few feet above the ground.

When hiking in remote places where you need your food for survival or where it's too difficult to replace any lost food immediately, protect it with a *double bagging system*. Instead of hanging all your food from one bag, put it into two bags and hang them 300 feet apart in opposite directions from your campsite. Although it's a lot more trouble to set up, you'll have greater peace of mind knowing that your food's a lot safer. Don't forget to carry two bags and two lengths of rope for this, though.

In popular camping areas like the national parks, bears have figured out that if they chew the rope tied to the tree, the whole sack of food will fall down. Prevent this by using the *counter-weight method* of hanging your food in a tree. Simply divide your smellable pile in half by weight and put each half in a separate sack (you'll need two sacks and one rope for this method). Throw the rope over a sturdy tree branch and tie it to one food bag according to the previous directions. Then, hoist that bag as high as it will go. Now tie the second bag to the rope and wrap any extra rope up with the bag so none hangs down below it. Then use a long stick to push the second bag up into the air so that it balances with the first bag about halfway down. Retrieve the bags in the morning by pushing one up with a very long stick until the other one lowers enough for you to reach.

You can double the security of any of the previously described food storage methods by securing the food sack in a tree with two ropes instead of one.

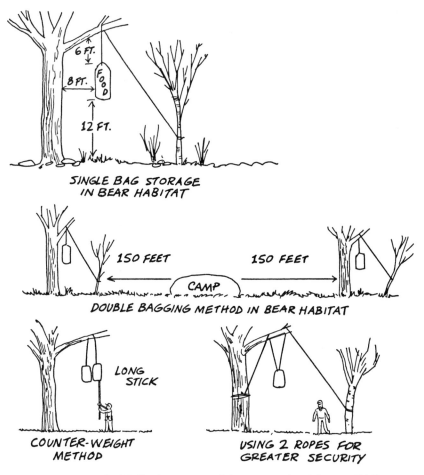

Example 20-3: Ways to safely secure food from wildlife.
Much shorter distances are satisfactory in areas without bears.

In treeless places like deserts, beaches, grasslands, and above timberline, rodents and not bears will be your main problem. Try placing your food on a large boulder with sides too steep for them to climb up.

In many popular camping areas, you should conceal your backpack (empty of all smellables) inside your tent because animals have learned that backpacks contain food and will assume there's food in your pack if they see it. When camping under a tarp, in a bivouac sack, or in the open, though, empty your pack, leave all its pockets open (unzipped), and keep it away from you when sleeping. That way, if an animal finds the pack, it can see there's no food in the opened pockets without tearing or chewing into it to investigate.

The last few paragraphs discussing food storage stressed securing food from bears in great detail because of the potential danger of a bear confrontation. By far your biggest food storage problem will be keeping squirrels, mice, and other rodents away from your food.

Animal Odds and Ends

Here are several things to remember when dealing with animals outdoors:

1) Every animal is more scared of you than you are of it.

2) Except for the dangerous animals that have lost their fear of people, all animals avoid contact with people as much as possible. All animals you meet, including the ones that have lost their natural fear of people, only want your food and want to be left alone. They don't want to attack you or confront you.

3) Avoid all strange behaving animals, especially if they approach you with no obvious signs of fear, because they could have rabies. Keep your tetanus shots current.

Snakes

There are four kinds of poisonous snakes in America. *Coral snakes* are small, secretive snakes averaging 2–3 feet in length found along the southeastern coastal plains and in the southern Arizona and New Mexico deserts. They have red, yellow, and black rings encircling their body with the red and yellow bands touching each other. Coral snake venom affects the human nervous system, which interrupts the operation of the body's major organs.

The three other kinds of poisonous snakes are called *pit vipers*. Their venom is injected by large fangs and attacks the human body's circulation system. The many colors and kinds of *rattlesnakes* which range over much of the United States are easily identified by the rattles on their tail. *Copperheads* are found throughout the east and midwest, average 3 feet in length, and have a copper coloring. *Water moccasins* live in the southeastern states along streams. They're also called cottonmouths because of the white coloring inside their mouths. You can identify the pit vipers by the physical characteristics described in Example 20-4.

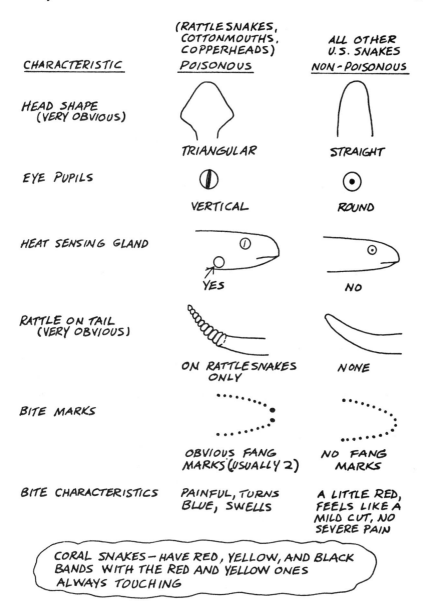

CHARACTERISTIC	(RATTLE SNAKES, COTTONMOUTHS, COPPERHEADS) POISONOUS	ALL OTHER U.S. SNAKES NON-POISONOUS
HEAD SHAPE (VERY OBVIOUS)	TRIANGULAR	STRAIGHT
EYE PUPILS	VERTICAL	ROUND
HEAT SENSING GLAND	YES	NO
RATTLE ON TAIL (VERY OBVIOUS)	ON RATTLE SNAKES ONLY	NONE
BITE MARKS	OBVIOUS FANG MARKS (USUALLY 2)	NO FANG MARKS
BITE CHARACTERISTICS	PAINFUL, TURNS BLUE, SWELLS	A LITTLE RED, FEELS LIKE A MILD CUT, NO SEVERE PAIN

CORAL SNAKES — HAVE RED, YELLOW, AND BLACK
BANDS WITH THE RED AND YELLOW ONES
ALWAYS TOUCHING

Example 20-4: General characteristics of poisonous
snakes in America.

Because snakes are cold-blooded animals, they take on the temperature of their surroundings. They can hardly move at all at temperatures below 45°F, seldom come out when the temperature is below 60°, and die when it gets over 100° for long periods of time. The temperature they like best is between 80–90°F. Note that those temperatures are measured at the ground level and not in the air. When the air temperature is 40° or 50°, the ground temperature could be above 80° and perfect for snakes. Look for them on sunlit rocks on cool spring and fall days, in the cooler mornings and evenings or in shade on hot days, and forget about them in winter or at high altitudes where it's too cold for them.

As a hiker, there are some interesting things you need to know about snakes. First of all, since their eyesight is very poor, they hunt by sensing heat and movement. Their tongue is used for smelling and not for eating or biting. They have rows of teeth inside their mouth they eat with. While rattlesnakes have rattles for warning possible enemies, they don't always rattle before striking, and snakes don't have to be coiled to strike. When coiled, they can strike up to two-thirds of their body length and can strike repeatedly at the same target.

It's not difficult to avoid snakes. Be careful when putting your hands and feet into crevices, under rocks, in woodpiles, and in bushes, because that's where they seek shelter. When hiking in a group, walk at least 10 feet behind the person in front of you, since, typically, the first person in a close group startles a snake and the second or third person is bitten. Use a flashlight when hiking at night so you don't walk on one by mistake. Wear long pants and stiff hiking boots for protection and walk with heavy footsteps, because snakes are sensitive to ground vibrations and seek shelter if they notice your presence. Freeze immediately if you hear a rattling or rustling sound near you because snakes frequently strike moving objects. If you see something moving into a bush near the trail, detour around the area instead of investigating to see what it was. In snake country don't go looking for trouble. Snakes will leave you alone if they are left alone.

A poisonous snakebite is not the danger it's often made out to be. There are only about 10-15 deaths from several thousand poisonous snakebites in America each year, and many of those people are bitten after handling pet snakes, working with snakes in zoos, or while working around their yard. Few hikers are ever bitten. The effects of a poisonous snakebite depends on your age, psychological condition, and amount of venom injected. It has a greater effect on people who are young, old, not in good physical condition, or emotionally unstable. Almost every adult of average build, good physical condition, and stable mental health can survive almost every poisonous snakebite. However, if you are bitten by a poisonous snake, see a doctor as soon as possible to be sure no permanent skin, tissue, or muscle damage results.

First Aid for Snakebites

If you're bitten by a snake, the first thing you should do is determine if it was poisonous or not by using the information on the chart on page 263, which you should memorize. If you're bitten by a *nonpoisonous snake*, simply wash the bite with soap and water and cover it with an adhesive bandage. No further medical help is needed unless the bite becomes infected. If you're bitten by a *poisonous snake*, quickly follow these steps in order:

1) Keep the bitten part of your body lower than your heart.

2) Stay calm, and don't panic. More people die from emotionally overreacting to the bite than from the snake venom itself.

3) Avoid physical activity, which increases your heart rate and pumps the venom through your system faster. Never take aspirin, alcohol, sedatives, or similar medications to ease the pain.

4) Evaluate the bite:

A *minor bite* has little or no local pain and swelling and no general body symptoms (see below). It means that very little or no venom was injected into your skin. More than one-third of all poisonous snakebites are of this type. You'll survive a minor bite with no complications. *Wash it with soap and water and check with a doctor as soon as you can, but without hurrying.*

Symptoms of a *moderate bite* include mild swelling and/or discoloration, mild to moderate local pain, nausea, shortness of breath, and rapid pulse. There are few general body symptoms. *You need to see a doctor as soon as possible to prevent medical complications, but your situation is not a critical life-or-death emergency.*

Symptoms of a *severe bite* include a piercing, burning pain; much rapid swelling; a throbbing, tingling, numbing sensation; and occasionally blood oozing from the bite. Generalized body symptoms include pinpoint pupils, slurred speech, paralysis, unconsciousness, a faster or slower heart rate, shortness of breath, overall weakness, poor vision, nausea, and vomiting. A severe bite indicates you received a lot of venom and the bite is quite serious. *With a severe bite you need to get to a doctor as soon as possible.*

And now the controversy begins. Doctors don't exactly agree on what is the best first aid for poisonous snakebites far from medical help. Some recommend cold applications, others recommend never using the cut-and-suck method described below, and still others recommend it without hesitation. The following information is compiled from a host of current books and articles as well as from the Red Cross Advanced First Aid text.

After you've decided you've received a moderate or severe bite from a poisonous snake, do some very fast thinking. If someone can carry you or if you can walk very slowly (with no backpack) to medical help, or if you can send someone who can return with skilled doctors within 4 hours or so, only do the first step listed below. If medical help is more than 4–6 hours away (depending on the severity of the bite), complete all the steps listed below.

5) Tie a loosely fitting *constriction band* 2–4 inches above the bite with a wide piece of cloth if it's on an arm or leg. It should be tight enough to contain the venom near the bite but loose enough so you can feel a pulse at the wrist or ankle and can slip a finger between it and the skin. Never tie a band around a joint like an elbow or knee or your head, neck, or waist.

6) If you're over 6 hours from medical help, make two shallow cuts ⅛-inch deep into fleshy skin through the fang marks and running for ½ inch vertically up the limb. Don't cut if there are visible arteries or veins just below the skin near the bite, and don't cut on the head, neck, or trunk areas of your body (Example 20-5). Make these cuts quickly since time is of the essence, but make them very carefully. Cutting improperly can do far more damage than snake venom.

7) Put suction cups from your snake bite kit on the cuts you made and suck out blood and venom. Suck with your mouth as a last resort if

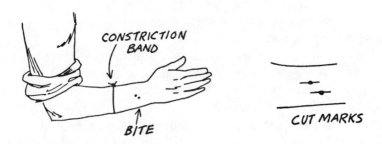

Example 20-5: First aid for severe poisonous snakebite far from medical help.

you have no suction cups, because poison in the wound could infect any cuts in your mouth. Up to one-half the injected venom can be removed if the cutting and sucking begin within 3 minutes of the bite, while little is removed if it begins 30 minutes after the bite occurred. Continue the suction for 20–30 minutes.

8) Rest in a comfortable place, keep the bitten part of your body as cool and as far below the level of your heart as possible, and mentally prepare yourself for a two-day period in which you'll feel terribly, painfully sick.

For general backpacking hikes in poisonous snake country, always carry a *snakebite kit* and know how to use it. When traveling for long periods of time far from medical help, carry an *antivenin kit* as well.

Killing snakes

The only time you should kill a poisonous snake is when people could be nearby and don't know where it is. Examples are when you find a poisonous snake at a trail shelter or when you see a poisonous snake along a trail and you know other hikers will be at that spot in a few minutes. Killing poisonous snakes at other times and killing nonpoisonous snakes at any time is a sign of your ignorance of the outdoors. May your soul bubble over with the spirit of life, not of violence and death. May your presence on this planet bring happiness, not terror, to others here. Live and let live.

Insects

Although *wasps, yellow jackets, honey bees, hornets,* and other *stinging insects* generally leave people alone, avoiding them will save you a lot of grief. Many of these insects leave their stinger in your skin, and the poison sac that's attached to it continues to pump poison after you are stung. Therefore, if the stinger is still in your skin, quickly and gently scrape it out with a knifeblade, stick, or fingernail. If you pull it out with your fingers or with tweezers, you'll squirt all the poison in the sack into your skin (Example 20-6). To neutralize the poison and reduce the pain, put meat tenderizer, ammonia, a baking powder paste, ice, or cold water on the stung area immediately. For most people in most cases, the pain and swelling rapidly disappear.

On rare occasions a victim stung by one of the stinging insects suffers *anaphylactic shock,* which is the body's violent reaction to the foreign proteins injected with the poison. Its symptoms include violent

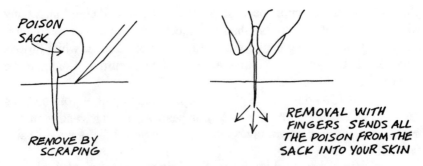

Example 20-6: Removing a bee stinger in your skin.

vomiting, nausea, cramps, and shortness of breath and can occur within several minutes to several days of the sting. Although it affects people known to be allergic to insect stings, it can even affect people who've been stung many times previously with no harmful effects. If a person shows signs of anaphylactic shock or was stung many times within a short period of time, keep him calm and take him to a hospital as soon as possible. Speed is essential, since there are no first aid measures you can treat anaphylactic shock with outdoors. When camping, always carry Benadryl or other medicine designed to neutralize poisonous stings if you know you're allergic to them.

There are many other insects you need to be aware of outdoors (Example 20-7). *Deer flies* are large flies whose bite is more painful than a mosquito but less than that of a bee sting. *Gnats* are smaller flies that have a less painful, more itchy bite. *Chiggers* are almost invisible bugs that cause the red dots and itchy feeling on your legs after walking through an overgrown field in short pants. *No-see-ums* are very tiny flies that can pass through mosquito netting and leave an occasional, painful

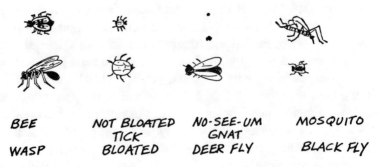

Example 20-7: Bothersome insects (drawn lifesize).

bite, but more frequently a general itchy feeling on your skin. *Black flies* are small flies with a bite that feels like a tiny pinch. There's little chance of disease or loss of blood from *mosquitoes*, but they leave an itching, annoying bite.

The bites from all those insects cause temporary symptoms and are almost always not dangerous. Protect yourself by wearing insect repellent, avoiding wet or swampy places where they breed, wearing long pants, long-sleeved shirts, and an insect head net where necessary, and camping in a floored tent with no-see-um netting (which is a finer mesh than regular mosquito netting).

Ticks are round, flying insects that painlessly attach themselves to your body to suck your blood. The only way to know they're on you is to thoroughly check your whole body, especially your scalp, several times a day. If you find one not bloated with blood, gently pull it out of your skin with a comb, tweezers, or your fingers. If it's attached firmly, touch its backside with a hot but not flaming match or pour a few drops of a thick oil like kerosene on it to smother it. Usually it'll let go to get air. Then, wash the area thoroughly with soap and water. If you can't remove a tick, if its head is still lodged in your skin after its body came out, of if the bite becomes red and infected, see a doctor immediately. Ticks occasionally carry Rocky Mountain Spotted Fever (which occurs all over America) and is easily cured if treated in time.

There are many other kinds of spiders and insects with poisonous bites. Generally, when bitten by any unknown insect or spider, cover the bite with ice or cold water and wash it with soap and water to prevent infection. Get to medical help as soon as possible if symptoms like headache, nausea, vomiting, or severe pain develop.

Desert Creatures

Scorpions hide in places like woodpiles, under rocks, and in old stumps during the day and hunt for insects at night. They kill their food and defend themselves from enemies by poisoning them with a stinger. Because the stings from several kinds of scorpions are extremely dangerous, always be alert for them when camping in the desert. Shake out your clothing and boots each morning to be sure no scorpions crawled in them and be careful when overturning rocks, collecting firewood, and climbing hillsides with bare hands. When stung, cool the area with ice water and get to a doctor immediately.

Desert centipedes grow to be 6-8 inches long, hide under logs or rocks in the day, and wander around hunting for insects at night. Their bite is usually a painful inconvenience rather than a serious injury. Wipe ammonia on the bite to lessen the pain, and see a doctor as soon as possible but without hurrying unnecessarily.

Tarantulas are large, hairy spiders with a painful bite which feels like and has the same effect as a bee sting. Wash the bite with soap and water and apply an antiseptic to prevent infection. People who try to pick them up are the only ones ever bitten, so leave these creatures alone.

Black widows are quarter-sized, entirely black spiders with a red hourglass shape on their undersides. They frequent woodpiles, outhouses, plants, and rocky crevices in the desert. Their bite is often painful and occasionally fatal. For first aid, cool the bite with ice, keep calm, and get to a hospital immediately.

Poisonous Plants

Poison ivy is a small vine or shrub that grows almost everywhere in the United States. Its leaves consist of three shiny green smooth, toothed, or lobed leaflets and often resemble "mittens" with the "thumbs" pointing outward.

Poison oak grows as a shrub or vine in many places in America. Its leaves consist of three hairy-bottomed leaflets with smooth, oval, or toothed edges. Like poison ivy, its leaves are arranged alternately on the stem.

Poison sumac grows as a woody shrub or small tree throughout the east. It has seven to 11 reddish leaflets, leaves arranged alternately on the stem, and white berries which distinguish it from the harmless red sumac.

To prevent getting an itchy rash from these plants, wash your skin and clothing with soap as soon after contact with them as possible. If you do get the rash, put calamine lotion or compresses soaked in a cold salt water solution on it. The itch usually disappears in a few days if you don't scratch it.

Nettles are plants that have small hairs that break off and inject an irritating substance into your skin when you touch them. Fortunately, the itchy feeling goes away after a while. Nettles are usually found in fields and thickets and have sawtoothed leaves.

Getting Help in an Emergency

When *hiking alone*, you'll have to solve your problems yourself. If you're near a trail, shelter, road, or town, get to it if possible so someone can help you. Use the signaling methods explained on page 281 to attract attention if you're injured so badly that you can't travel. Hopefully you told someone where you were going and they'll send a search party for you when you don't return on time.

When *hiking in a group of two people*, the unhurt person will probably have to leave the victim to get help. Before leaving, though, be sure you know where you're going and carry sufficient supplies like water, food, map, flashlight, and clothing to get you there. Carry money for a phone call, shelter if going overnight, and a first aid kit for emergencies. Tell or leave a message with the victim which way you're going so people can find you if you run into trouble. Before leaving your partner, draw a map showing his exact location, and mark his location with a bright piece of clothing or emergency signal so others passing by can help him if they find him before you return. If there's a chance you'll be gone overnight or in very cold weather, leave plenty of firewood, warm clothing, shelter, food, and water with the victim as well.

In *groups of three people*, one person should go for help and one should stay with the victim. In *groups of four or more*, two people should go for help and everyone else should stay with the victim.

A fast way to travel in an emergency is to alternate 50 running steps with 50 walking steps. You can find a comfortable pace and travel for hours this way if necessary.

21

Survival

Lost

Few hikers are truly lost outdoors. Often they're simply disoriented because they took the wrong trail or misread their map. In most cases, it's relatively easy to orient yourself again and become "unlost."

As soon as you think you're lost, stop, think, and rest. Listen for any familiar sounds like a stream, a highway, or a train. Climb a tree or nearby hill to see if you can figure out where you came from or where you should be. Compare obvious landmarks with those on your map. Don't keep on walking or wandering around, because you could get yourself more lost. Don't take any rash actions like running through the woods or hiking to a distant mountain to see where you are. Above all, don't panic. Chances are you're only a short distance from where you should be.

Draw a sketch map on paper or dirt showing your present location, any prominent landmarks, the compass directions, and where you think you last knew your exact location. Mark any rivers you crossed, any mountains or ridges you climbed, and any overlooks or viewpoints you passed. Mark the times (even measured in terms of "a short distance" or "a long way") between any markings on the map and mark if the trail was generally flat, mostly uphill, or mostly downhill. Draw all *baselines*, like roads, rivers, or mountain ranges on the map, even if they're 20 miles away. They'll give you reference points to put your own position in perspective. Then compare your sketch map with your topographic map to help locate your position.

If you're sure you're not far from where you should be, mark your present location with an object like a brightly colored shirt tied to a tree.

Then blaze a trail with sheets of toilet paper impaled on tree branches or other obvious markings in the direction you believe you need to go. Constantly look over your shoulder as you walk to memorize the way if you have to return again. If you find the place where you got lost or a place where you know the location, have your group wait there while someone retraces their steps to remove the toilet paper markers and retrieve the shirt (be sure to walk back to the shirt first and then remove the paper so you don't get lost walking back). If you hiked a fair distance and still aren't sure of your location, retrace your steps along the toilet-paper-marked trail (removing it as you go) and stay near the spot marked by the shirt.

If you're still lost, assess your situation. Determine how much daylight is remaining and how much food, water, shelter, and equipment you have. You'll seldom be lost with nothing useful at all. Check your pockets, backpack, and surroundings for anything that could help you survive until someone finds you. Conserve your strength, get water if needed, collect firewood, and build some kind of shelter. Maintain group morale, don't become upset, and follow the specific suggestions for survival later in this chapter. When lost, it's better to wait for rescuers to find you than to wander around aimlessly trying to become "unlost," assuming again, that someone knows where you are and when you planned to return.

Getting "Unlost"

If you're lost and no one knows where you are, you'll have to get back to civilization yourself or stay where you are and signal for help (see page 281). Never travel "back to civilization" unless you're thoroughly physically and mentally prepared for traveling under potentially severe conditions. Obviously, if you're sick, injured, lost in a very remote area, or stranded in bad weather, it's better to remain in one place and signal for help than to try to walk out.

Before walking anywhere, be sure you're going somewhere. Find any baselines on your topographic map or on the sketch map you drew earlier, especially if they're roads, towns, railroad lines, or trails which will eventually lead to inhabited areas. In relatively inhabited areas, streams are good baselines because they usually lead to farms, villages, or roads, but wilderness streams often end up at obstacles like waterfalls, swamps, or lakes, or simply lead farther from civilization. Plan to walk the easiest, most direct cross country route in as straight a line as possible to the nearest baseline leading to civilization. Use your compass as a guide to help you walk in that straight line (see "Traveling in a Straight Line" below) to civilization and not as your only guide leading you blindly onward to an unknown destination. Walking north for the sake of walking north will only get you farther north and possibly farther lost.

Then, after you know exactly where you're going, prepare for your journey by carefully packing necessities including water, map, compass, food, clothing, shelter, matches, rope, flashlight, and first aid kit, while leaving anything not absolutely needed like books, garbage, and camp moccasins behind to lighten your pack. As you travel make a sketch map which should include your approximate hiking times and distances between noticeable landmarks and topographical features. Mark your trail with pieces of tissue paper or obvious markings in case you reach an obstacle and have to backtrack.

When lost with a group of people, your options increase significantly. Depending on your particular situation and circumstances, including the group morale, the amount of each individual's outdoor experience, the amount and type of available resources, the terrain, the prevailing weather conditions, and the urgency of the situation (any injured? impending storms? lack of water?), you can all hike out or all stay put and signal for help as one unit or split into two groups with one group staying in one place and another smaller, better equipped, and more experienced group walking out for help.

Traveling in a Straight Line

Usually the fastest way to walk to help is to walk in a straight line. Several ways to do that accurately are listed below:

1) In flat, open country, simply pick out a distant mountain or landmark and walk towards it.

2) Use the sun as a guide if you know which direction you need to walk. Remember that generally the sun rises in the east, is south at noon, and sets in the west. When walking north, keep it on your right in the morning, behind you at midday, and on your left in the evening.

The following methods are much more accurate than the previous two:

3) *Straight line walk:* First figure out the direction you want to walk. Then line up two trees, rocks, or other distant, prominent objects in a straight line in that direction and walk to the closer of those objects.

Before reaching it, line up those first two objects with a third object and walk to the second one. Before reaching the second object, line it up with the third object and a fourth object and walk to the third one. Repeat as long as necessary (Example 21-1).

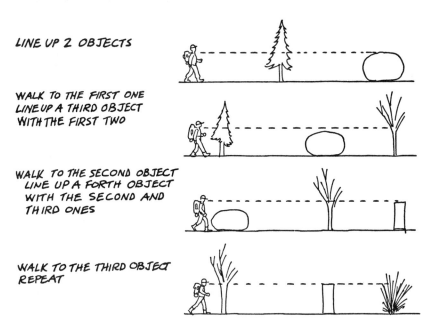

LINE UP 2 OBJECTS

WALK TO THE FIRST ONE
LINE UP A THIRD OBJECT
WITH THE FIRST TWO

WALK TO THE SECOND OBJECT
LINE UP A FORTH OBJECT
WITH THE SECOND AND
THIRD ONES

WALK TO THE THIRD OBJECT
REPEAT

Example 21-1: Straight line walk.

4) *Compass bearing walk:* Take a bearing with your compass on a distant object in the direction you want to travel. Then put the compass away, walk to the object, and take another bearing to a farther object from there. Put the compass away again, walk to the next object, and take another bearing to a still farther object. Repeat as often as necessary (Example 21-2).

Walking Around an Obstacle

If you're following a *straight line with no compass* and suddenly find an obstacle like a lake or swamp in your path, pick out two prominent landmarks in line with your course on its distant shore and detour around the obstacle until you reach your first landmark. Continue using the straight line walk being sure to always line up a third object with two closer objects before continuing (Example 21-3).

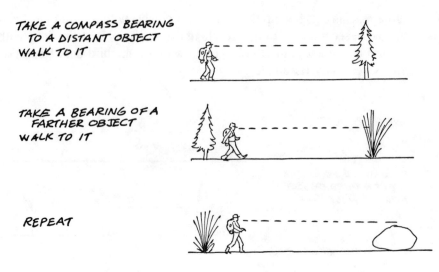

Example 21-2: Compass walk.

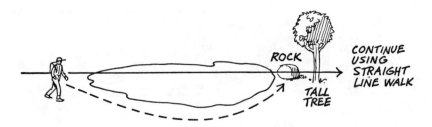

Example 21-3: Walking around an obstacle with no compass.

Traveling around an obstacle like a lake or swamp *with your compass* is simple if you can spot a landmark like a tall tree or a huge rock on its opposite shore that's on the bearing you're following. Just detour around the obstacle until you reach your landmark and measure the bearing with your compass from there (Example 21-4).

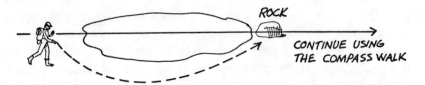

Example 21-4: Walking around an obstacle with a compass.

If you can't see across an obstacle like a cliff, you can still keep your bearing by walking around it using a *right angle walk* (Example 21-5). A compass adds a great deal of accuracy to this method, but it's not absolutely necessary for it to work. Follow these steps:

1) When you reach the obstacle, add 90° to your bearing and turn to your right.

2) Count the number of steps you take as you follow the new bearing until you clear the obstacle. Try to keep the size of your footsteps as constant as possible throughout this exercise for greater accuracy.

3) Now subtract 90° from your bearing (you'll have the first bearing again), and follow that until you're sure you cleared the obstacle.

4) Subtract 90° again, turn to the left, and walk the same number of steps as you did earlier.

5) Add 90°, turn to the right, and follow your original bearing again.

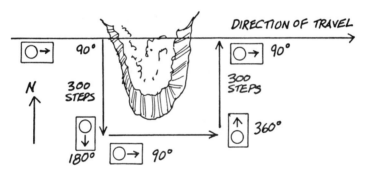

Example 21-5: Detouring around a large obstacle like a cliff with a right angle walk.

If you want to detour around an obstacle to the left instead of to the right, simply add where the above directions said to subtract and subtract where they said to add.

When using the method illustrated in Example 21-5, you can often save a great deal of time and mileage if you make a triangular and not a square detour around the obstacle (Example 21-5a). You can vary the size of the triangle to optimize your detour time. In order to return to your original bearing when using this method, just be sure that angle 1 and angle 2 are equal, that the sum of all three angles equals 180°, and that the number of steps you take on detour 1 is identical to the number of steps you take on detour 2.

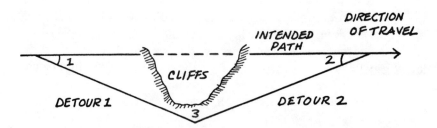

Example 21-5a: Detouring around an obstacle with the triangle method.

Staying Alive

While the skills of hunting wild game and building lean-to shelters has without a doubt saved people lost in the wilderness, and while those survival skills might still be applicable in remote backcountry areas, they've become just an entertaining pastime for modern-day armchair pioneers who camp in most areas in the continental United States. In most places in America, knowing how to snare a rabbit in a survival situation is useless, for survival has come to mean not sustained existence in the wilds but patient endurance until rescuers arrive. Read one of a host of survival books on the market if you think you'll need to know how to live off the land. The next several pages stress practical aspects of survival when lost or stranded outdoors that'll help you stay alive long enough to get rescued.

Be Prepared

Survival begins at home. Check the weather report so you know what kind of weather to expect and pack your gear accordingly. Use common sense while hiking. Don't push yourself too far, and carry extra clothes, food, and water. Practice the map reading, firebuilding, and campcraft skills you'll need on an easy, undemanding weekend outing before relying on them under trying circumstances. Know and accept the risks of hiking alone. Finally, tell someone where you're going and when you'll return. The sooner someone knows you are gone, the sooner they'll come looking for you.

Food

Generally, when lost or stranded, stay in one place, conserve your energy, and wait to be found. Hunting wild game and looking for edible plants are for Robinson Crusoe types lost in the wilderness with no chance of being found. By conserving your energy, you can survive for over one month with no food, while you can literally starve by wandering around looking for edible wild foods that are hard to find and difficult to catch.

Water

When water is scarce, get out of the wind and the sun. Don't sweat and don't breathe heavily. Never drink sea water, urine, or water from bad springs, no matter how thristy you are, because they contain high concentrations of salt which makes you more thirsty and unbalances your body's natural salt equilibrium. Never eat food (except juicy fruits) when water is scarce because the process of digestion requires more water than the foods provide.

You can't live for more than one or two days without water so carefully conserve what you have and don't wait to search for more until you have none left. In wet climates look for streams or springs in gullies and in valleys. In deserts look for rainwater trapped in rocky potholes in shady canyons and springs near cottonwood trees. Although many survival books recommend obtaining water by chopping open a barrel cactus, by digging holes in the ground, and by mopping up dew with your clothes, constructing a *solar still* is the most dependable method of obtaining drinking water that requires little effort on your part and is effective almost anywhere (Example 21-6).

To construct a solar still, you'll need some type of container and a sheet of plastic. A piece of plastic tubing is optional. For best results, the

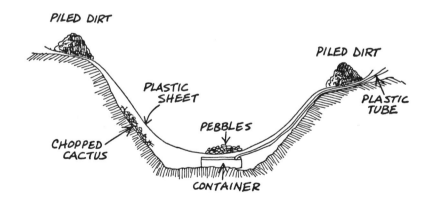

Example 21-6: A solar still.

container should be wide, large, and plastic. Small containers don't catch all the water produced, while metal ones heat up and evaporate some of the water in them. In an emergency, though, anything like a can, cooking pot, or hubcap will do. The plastic sheet should be transparent and at least 4–5 feet long on each side, although a ground sheet the size of a sleeping bag or a black plastic garbage bag works in an emergency. Patch all holes in the plastic sheet with tape, and lightly scrape the bottom surface of it with sand so water drops can adhere to it better.

Construct the still in a sandy wash, a grassy meadow, or a depression in the ground which holds rainwater longer than the surrounding area, and construct it so it gets as much sunlight as possible throughout the day. Dig a hole about 3 feet in diameter and a little less than 2 feet deep, depending on the size of the plastic sheet you use. Put the container in the bottom center of the hole, stick one end of the plastic tube in the bottom of the container, and run the other end out of the hole. Next, lay the plastic sheet over the hole so that it's about 2–4 inches above but never touching the ground inside the hole and along its sides. Weigh the center down with a few pebbles to make a cone that takes the shape of the hole. Then pack dirt around the outside edges of the plastic to seal out air. While a properly constructed solar still yields from 1–2 quarts of water a day, you can improve its effectiveness by lining the sides of the hole with green leaves, chopped-up cacti, polluted water, or body wastes before covering it with the plastic sheet.

In theory the sun's rays evaporate water from the soil inside the still, which condenses on the underside of the plastic and drains into the container. The still will begin producing water 1–2 hours after you finish constructing it. To drink the water, simply sip on the plastic tube or remove the soil and plastic to retrieve the container of water. You'll lose some water production time though, every time you take the still apart to retrieve the container. For continued results, move the still to a new location every two to four days.

If you hike extensively in dry places like deserts, its a good idea to buy and carry a *solar still kit* designed specifically for evaporating drinking water. They are more efficient and reliable than a plastic ground cloth used in an emergency.

Fire

While some survival books claim that you can build a fire with ice, eyeglass lenses, flint and steel, flashlight batteries, or a bow and drill, those methods work under ideal conditions when you're warm, happy, and experienced with them. When you're cold, wet, and hungry and its dark and you really need a fire quickly, you'll never be able to light one with those methods. Carry dry matches, a pocket knife, and a candle stub or firestarter, and with some patience, you'll be able to light a fire anywhere under almost any conditions, by using the techniques explained on pages 207-209.

Shelter

If you're lost, the best shelter you can use is your own tent, tarp, poncho, or ground sheet which you should always carry. If for some reason or disaster you're without those things, find a protected place out of the snow, rain, and wind and build a shelter that's as simple, easy to construct, and weatherproof as possible.

In a forest pile sticks, branches, and logs against downed trees. Uprooted evergreen trees are best, because their thick branches are naturally protective. In rocky country, seek shelter in narrow clefts in cliffs, under boulders, or in a gully covered with logs and branches. Insulate your shelter by covering its roof and walls with green tree boughs, chinking cracks between rocks with moss or pebbles, and piling dry leaves, grass, or pine needles on the ground inside it. Fill your pockets and the spaces between clothing layers with dry leaves, grass, or pine needles for more insulation on cold nights.

Signaling for Help

When lost or stranded, it's often better to stay in one place and signal for help than to try to find your way back to civilization. Of course, if you didn't tell anyone where you were going and when you expected to return, you'll either have to signal for help or travel to civilization yourself. Signaling for help is a lot more effective when people are actively looking for your signals.

When rescuers know what trail you were on or your general location, they'll often search for you from the ground. When they don't know where you are, they'll run a spot search by airplane first. Because search planes fly a regular pattern over areas where they think you are only once until they've completed the whole territory they're searching, it's very important not to miss that search plane when it passes over. Since you'll never know when the search plane will fly over or which passing plane is a search plane looking for you, you have to be constantly ready to signal every plane flying nearby.

Signals must be different to be noticed easily. Use sounds, colors, and patterns that *contrast sharply* with their backgrounds. Build a bright, flaming fire at night and a smoky fire during the day. Spread colorful clothes and sleeping bags out on drab-colored backgrounds. Hang colorful balloons from trees so that they stand out to passersby and airplanes. Tie clothing in trees. Write a large "SOS" on the ground or in the snow with your footprints and make the letters stand out by lining them with green tree branches, logs, or dirt.

Anything in *groups of three* is a signal for help. Blow three times on your whistle and then wait for any replies. Build three fires about 20 feet apart in a straight line or in the shape of a triangle. Shine your flashlight in flashes of three.

The letters *SOS* mean distress. Stamp them in the snow, write them in a meadow with logs or rocks, or write them on a nearby trail with pebbles.

Use a *fire* but use it wisely. Don't burn down a whole forest while trying to be rescued. Keep a small fire burning all the time, have piles of tinder and kindling ready to build it up quickly, and have wet leaves or green brush ready to throw onto the fire to make smoke when a plane flies overhead.

To use a *signaling mirror* (a double-sided, metal mirror with a hole drilled in its center), simply look through the hole at a lookout tower, airplane, or similar object you want to contact. Now, with your eyes still looking through the hole at the object, turn the mirror so that the sun shines through the hole and leaves a white dot of light on your face (you'll see it on the back side of the mirror). Turn the mirror until the white dot disappears into the hole of the mirror, which indicates the sun's light is reflecting to the object you're looking at. Wag the mirror back and forth for added effect. Pilots pay much more attention to a flashing reflection than a stationary one. Use two fingers in the shape of a "V" as a guide when aiming at a moving object like a plane (Example 21-7).

The method just described works best when the object and the sun are less than 90° apart or when the sun is high in the sky. It's much more difficult to signal with a mirror when the sun is near the horizon and the object you want to contact is more than 90° from it. Practice using a signaling mirror at home first to increase your speed and accuracy with it.

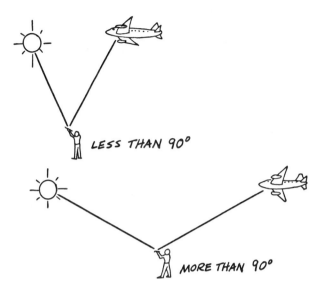

LESS THAN 90°

MORE THAN 90°

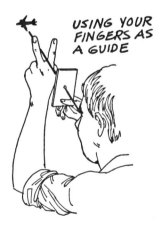

USING YOUR
FINGERS AS
A GUIDE

Example 21-7: Using a signaling mirror.

V
APPENDIX

APPENDIX—
Sources of Information

The following lists of addresses are a representative sampling of companies and organizations serving backpackers. They are by no means complete or an indication of the quality of the products or organizations.

Catalog Equipment Suppliers

The following companies market products directly to consumers through catalogs, which are available upon request. As a general rule, a catalog should clearly state the following information:

... the weight of all items listed,
... the company's guarantee and return policy,
... who pays for shipping and handling charges,
... the exact shipping procedure (UPS, Parcel Post, etc.),
... how to measure yourself to fit their sizes,
... the kind and quality of materials used.

In addition, better companies offer prices comparable to those in local stores and practical, useful products that aren't simply showcase fashions or eye-catching gimmicks. The chart on page 289 accompanies the following list of addresses and tells you what kind of gear they supply.

1) Campmor
   810 Route 17
   Paramus, NJ 07652
2) Don Gleason Camping Supply
   Pearl Street
   Northampton, MA 01060

3) Early Winters
   110 Prefontaine Place S.
   Seattle, WA 98104
4) Eastern Mountain Sports
   2410 Vose Farm Road
   Peterborough, NH 03458

5) Eddie Bauer
Fifth and Union, Box 3700
Seattle, WA 98124

6) Frostline Kits
Frostline Circle
Denver, CO 80241

7) Indiana Camp Supply
Box 344-SJ2
Pittsboro, IN 46167

8) L.L. Bean
3661 Main Street
Freeport, ME 04033

9) Ramsey Outdoor Stores
226 N. Route 17
Paramus, NJ 07652

10) Recreation Equipment Inc.
Box C-88125
Seattle, WA 98188

11) Yak Works
2030 Westlake Ave.
Seattle WA 98121

Equipment Manufacturers

The following companies manufacture various kinds of back-packing gear. Most will mail you a free catalog upon request. Some of these companies sell their products through catalogs and through local retailers, while others sell them only through local outdoor stores.

a) Alpenlite
3891 N. Ventura Ave.
Ventura, CA 93001

b) Bibler Tents
1715 15th St.
Boulder, CO 08302

c) Black Ice
2310 Laurel St.
Napa, CA 94559

d) Camp 7
1275 Sherman
Longmont, CO 80501

e) Camp Trails—Johnson Camping
Box 966
Binghamton, NY 13902

f) Coleman Company
250 N. Saint Francis
Wichita, Kansas 67021

g) Diamond Brand
Naples, NC 28760

h) Eureka Tents
625 Conklin Road, Box 966
Binghamton, NY 13902

i) Jan Sport
Paine Field Industrial Park
Building 306
Everett, WA 98204

j) Kelty Pack Company
Box 639
Sun Valley, CA 91352

k) Kirkham's Outdoor Products
3125 S. State St.
Salt Lake City, UT 84115

l) Lowe Alpine Systems
Box 189
Lafayette, CO 80026

m) Marmot Mountain Works
331 S. 13 St.
Grand Junction, CO 81501

n) Moss Tents
Mt. Battie St.
Camden, ME 04843

o) Mountain Safety Research
P.O. Box 3978
Terminal Station
Seattle, WA 98124

p) North Face
1234 Fifth Ave.
Berkeley, CA 94710

q) Optimus Stoves
P. O. Box 1950
1100 Boston Ave.
Bridgeport, CT 06601

r) Outdoor Products
533 S. Los Angeles St.
Los Angeles, CA 90013

s) Patagonia Clothing
Box 150, Dept. FF
Ventura, CA 93002

t) Sierra Designs
247 Fourth St.
Oakland, CA 94607

u) Sierra West
6 East Yanomali St.
Santa Barbara, CA 93101

v) Silva Compasses
P. O. Box 966
Binghamton, NY 13902

w) Stephenson Outdoor Products
Dept. 0
Gilford, NH 03246

x) Wilderness Experience
20675 Nordhoff St.
Chadsworth, CA 91311

The information presented here is based on recent catalogs available from the addresses on the previous pages.

Catalog Suppliers:

Number from previous page	Packs	Sleeping Bags	Shelters	Boots	Clothing	Gadgets	Cook Gear	Food	Books	Climbing Gear	Cycling Gear	Boating Gear	Cross-Country Ski Gear
1	X	X	X	X	X	X	X		X		X		
2	X	X	X		X	X	X			X	X		X
3	X	X	X	X	X	X	X	X			X		X
4	X	X	X	X	X	X	X	X	X	X	X	X	X
5		X		X	X								
6	X	X	X		X						X		
7	X	X	X		X	X	X	X	X				
8	X	X	X	X	X	X	X				X		
9	X	X	X	X	X	X	X						
10	X	X	X	X	X	X	X	X	X	X	X	X	X
11	X		X	X	X	X	X				X		

Manufacturers:

Number from previous page	Packs	Sleeping Bags	Shelters	Boots	Clothing	Gadgets	Cook Gear	Food	Books	Climbing Gear	Cycling Gear	Boating Gear	Cross-Country Ski Gear
a	X												
b			X										
c		X	X		X								
d		X	X		X								
e	X												
f	X	X	X				X					X	
g	X		X										
h			X										
i	X		X		X								
j	X	X	X		X								
k	X	X	X				X						
l	X	X		X	X								
m		X	X		X								
n			X										
o					X		X			X	X		
p	X	X	X		X								
q							X						
r	X	X			X								
s					X								
t			X										
u			X										
v						X							
w		X	X										
x	X	X	X		X								

Lightweight Hiking Boot Suppliers

The following is a list of manufacturers and suppliers of lightweight hiking boots. Most will send you a catalog or information sheet upon request. Note that many general catalog suppliers listed on page 289 also market hiking boots.

1) Asolo Sport
   Kenko International
   8141 W I-70 Frontage Road N.
   Arvada, CO 80002

2) Danner Shoe Company
   P.O. Box 22204B
   Portland, OR 97222

3) Donner Shoe Company
   2110 Fifth Street
   Berkeley, CA 94710

4) Fabiano Shoe Company
   850 Sumner Street
   South Boston, MA 02027

5) Inter Footwear USA
   4500 (A) North Star Way
   Modesto, CA 95356

6) Jung Shoe Company
   P.O. Box 28
   Sheboygan, WIS 53081

7) Kastinger Boots
   Brenco International
   7835 S. 180th St.
   Kent, WA 98032

8) Merrell Boot Company
   Hiddenwood/German Flats Road
   Route 1, box 105
   Waitsfield, VT 05673

9) New Balance Shoes
   38-42 Everett St.
   Boston, MA 02134

10) Nike Shoes
    3900 S.W. Murray Blvd.
    Beaverton, OR 97005

11) Pacific Mountain Sports
    La Canada, CA 91011

12) Rocky Boots
    P.O. Box A
    Nelsonville, OH 45764

13) Vasque Boots
    Red Wing Shoe Company
    Red Wing, MN 55066

14) Wolverine Boots and Shoes
    Rockford, MI 49351

Backpacking Food Distributors

The following list supplements the previously listed general catalog suppliers that supply backpacking foods. Most will send you a catalog upon request.

1) Camplite
   Infinity Nutritional Foods
   P.O. Box 391
   Montclair, CA 91763

2) Chuck Wagon Foods
   Micro Drive
   Woburn, MA 01801

3) Mountain House Foods
   Oregon Freeze Dry Foods
   Box 1048
   Albany, OR 97321
4) Rich Moor Corp.
   Box 2728
   Van Nuys, CA 91404

5) Seidel's National Trail Foods
   18607 Saint Clair Ave.
   Cleveland, OH 44110
6) Stow-Lite Foods
   Stow-a-Way Sports Industries
   166 Cushing Highway, Route 3A
   Cohasset, MA 02025

Materials

1) W. L. Gore and Associates   Gore Tex™
   Box 1220
   Elkton, MD 21921
2) DuPont Company   Hollofill™
   Fiberfill Marketing   Sontique™
   Centre Road Building
   Wilmington, DE 19898
3) Howe and Bainbridge, Inc.   Klimate™
   220 Commercial Street
   Boston, MA 02109
4) Reliance Products   Polarguard™
   1614 Campbell
   Oakland, CA 94607
5) Helly-Hansen, Inc.   polypropylene
   P.O. Box C-31
   Redmond, WA 98052
6) 3M Company   Thinsulate™
   223-6SW
   3M Center
   Saint Paul, MN 55114
7) Woolrich Company   wool
   Box B-81   (ask for the free booklet
   Woolrich, PA 17779   "Information That Can Save
   Your Life")   Your Life")

Outdoor Magazines

1) *Backpacker Magazine*
   P.O. Box 2784
   Boulder CO 80321

2) *Outside Magazine*
   P.O. Box 2690
   Boulder, CO 80321

Organizations

Government

1) Bureau of Land Management
   Office of Information
   Department of the Interior
   Washington, DC 20240
2) National Park Service
   Department of the Interior
   Washington, DC 20240
3) U.S. Fish and Wildlife Service
   Department of the Interior
   Washington, DC 20240
4) U.S. Forest Service
   Department of Agriculture
   Washington, DC 20250

Conservation/Private

1) Adirondack Mountain Club
   Gabriels, NY 12939
2) American Hiking Society
   317 Pennsylvania Ave.
   Washington, DC 20003
3) American Youth Hostels
   132 Sprint Street
   New York, NY 10012
4) Appalachian Mountain Club
   5 Joy Street
   Boston, MA 02108
5) Appalachian Trail Conference
   Box 236
   Harpers Ferry, W VA 25425
6) Boy Scouts of America
   P.O. Box 61030
   Dallas/Fort Worth Airport, TX 75261
7) Colorado Outdoor Center for the Handicapped
   Box 697
   Breckenridge, CO 80424
8) Earth First!
   Box 235
   Ely Nevada 09301

Organizations (cont.)

9) Friends of the River
   Box 1115
   Flagstaff, AZ 86002

10) Girl Scouts of America
    830 Third Ave.
    New York, NY 10022

11) GreenPeace
    240 Fort Mason
    San Francisco, CA 94123

12) Federation of Western Outdoor Clubs
    16603 53 Ave. S.
    Seattle, WA 98188

13) National Audubon Society
    950 Third Ave
    New York, NY 10022

14) National Hiking and Ski Touring Association
    Box 7421
    Colorado Springs, CO 80907

15) National Outdoor Leadership School
    Department G, Box AA
    Lander, WY 82520

16) National Wildlife Federation
    1412 16th Street NW
    Washington, DC 20036

17) Outward Bound, Colorado
    945 Pennsylvania Street
    Denver, CO 80203

18) Sierra Club
    1050 Mills Tower
    San Francisco, CA 94104

19) U.S. Orienteering Federation
    P.O. Box 1039
    Ballwin, MO 63011

20) The Wilderness Society
    1901 Pennsylvania Ave. NW
    Washington, DC 20006

A flaming sunset is such a spectacular sight
As it blends into the starlit twilight
Vega begins to shine forth like a guiding light
And soon a heavenly bonfire will burn in the night

I can see the world changing from this height
I can see suburbs growing like a terrible blight
People gather houses, cars, and things with all their might
Their whole world is drifting and spinning with fright

And now the future's coming and it's revealing my plight
I can see my values changing and myself refusing to fight
Maybe the wind will guide me in my flight
Maybe books of wisdom will help me see wrong from right
Or maybe I just gotta climb every mountain to keep my
*   carefree spirit burning bright . . .*

Index